I0485858

Exploring Pilots' Experiences of Integrating Technologically Advanced Aircraft Within General Aviation

Calvin Nobles, Ph.D.

Author:	**Calvin Nobles, Ph.D.** Cn8972@gmail.com
Publisher:	**Dawn D. Boyer, Ph.D., D. Boyer Consulting** **Virginia Beach, VA 23464;** Dawn.Boyer@me.com
Copyright ©	2015
ISBN Numbers	ISBN-13: 978-1517039943 ISBN-10: 1517039940

All rights reserved. No part of this book, including interior design, cover design, icons, and pictures, may be reproduced or transmitted in any form, by any means (electronic, photocopying, recording, or otherwse) without the prior permission of the copyright owner.

Trademarks:	All brand names, product names, logos, service marks, trademarks or registered trademarks are trademarks of their respective owners. All photos in the book were labeled "Public Domain, Permission to Reuse, with or without attribution to photographers" from sources. All attempts to note full name of photographers were made when and where possible.
Cover Courtesy of / copyrighted by:	Designed by Dawn D. Boyer, Ph.D., Copyrighted 2015©
Title Page Photo:	Tailwind 61E8X is a W-10, built by Jim Clement of Baraboo WI USA. Photo from EAA convention site, Oshkosh WI USA; Source: Wikpedia Public Domain photographer FlugKerl2.

Disclaimer:
The author has attempted to gather as much of the facts and information to the utmost complete and truthfulness for the compilation of this book from bona fide sources, internet sources, printed material in currently circulating and non-circulating sources, newspaper articles, personal interviews. Dates noted were from publically available sources.

Keep in mind – if any data included (or left out, incorrectly quoted, or attributed), it may be attributed to transcription errors or types. Several bodies of research were interpretations of the same or original documents and errors might have occurred as transcribed. Anyone with more data to contribute to a future, updated, and corrected version of this project is encouraged to send materials to the author directly.

Connect to the Author

LinkedIn profile: https://jp.linkedin.com/in/calvinnobles

Exploring Pilots' Experiences of Integrating Technologically Advanced Aircraft Within General Aviation: A Case Study

by

Calvin Nobles, Ph.D.

2015, Doctor of Philosophy (Ph.D.), Business Administration, Northcentral University, Prescott, AZ
2010, Master's of Military Operational Art and Science, Air Command and Staff College, Maxwell AFB, AL
2009, Master's of Business Administration, Northcentral University, Prescott, AZ
2004, Master 's of Aeronautical Science, Embry Riddle Aeronautical University, Daytona Beach, FL
1998, Bachelor of Science, Management, Park University, Parksville, MO

A Dissertation Submitted to the Faculty of
Northcentral University Graduate Faculty of the
School of Business and Technology
in Partial Fulfillment of the
Requirements for the Degree of

DOCTOR OF PHILOSOPHY
January 2015

Approved by:
Chair: Dr. Robert George, Ph.D.
Member: Dr. Martin Crossland, Ph.D.
Member: Dr. Nicole Avena, Ph.D.
Dean of School: Dr. Peter Bemski, Ph.D.

This text has been altered in format from the original dissertation document to conform to easier readability for the general public and commercial publishing standards. Scholars reviewing the contents and formatting for standardization for thesis or dissertation should not use this book's current formatting as a model. Please see your school's guidelines for the prescribed and acceptable formatting for the graduate level thesis researched for your degree publication – or – reference a professional academic editor for guidance.

Copyright, 2014, by Calvin Nobles

All rights reserved worldwide. No part of this book, including interior design, cover design, icons, and pictures, may be reproduced or transmitted in any form, by any means (electronic, photocopying, recording, or otherwise) without the prior permission of the copyright owner. Independent of the author's economic rights, and even after the transfer of the said rights, the author shall have the right to claim authorship of the work and to object to any distortion, modification of, or other derogatory action in relation to the said work, which would be prejudicial to the author's honor or reputation.

TABLE OF CONTENTS

List of Figures 8
List of Tables...10
Acknowledgements & Forward 11
Abstract...12
CHAPTER I - Introduction 17
Background 21
Statement of the Problem.............................27
Purpose of the Study...................................29
Theoretical Framework....................................33
Research Questions....................................49
Nature of the Study.....................................50
Significance of the Study 56
Definition of Key Terms 59
Summary 67
CHAPTER II - Review Of Literature 71
General Aviation Initiatives...........................102
Leveraging Technology to Improve Safety.................109
Determining the Effects on Human Factors...............119
Improving Flight Training to Address New Technologies
....136
Capitalizing on Innovations by Reducing the Negative
Influences 145
Management Practices in General Aviation 151
Summary 174
CHAPTER III - Methods And Procedures............177
Research Methods and Design 182
Population..198
Sample 200
Materials/Instruments 203
Instrument pilot test.......................................207

Data Collection, Processing, and Analysis 210
  Determining initial contacts/Screening. 210
  Consent 211
  Enrollment 212
Assumptions 227
Limitations 228
Delimitations 230
Ethical Assurances 230
Summary 233
CHAPTER IV - Findings 235
Results 237
  Critical Findings. 242
  Private Pilots Demographic Data 244
  Results 249
  Major themes of the study. 254
  Private Pilot (PP) Group. 260
  Commercial-Instrument Rated (CIR) Pilot Group. 270
  Certified Flight Instructor (CFI) Group. 282
  Comparison of the Embedded Units of Analysis. 292
Evaluation of Findings 294
  Evaluation of findings for Research Question (RQ) 1 295
  Regulatory 297
  Training 299
  Challenging and Transformative Transition. 300
  Evaluation of findings for RQ2 302
  Situational Awareness. 302
  Training. 304
  Aeronautical decision-making and Single Pilot Resource
  Management. 306
  Evaluation of findings for RQ3 309
  Training. 309
  Proficiency and Familiarization. 311
  Safety. 314
  Evaluation of findings for RQ4 316
  Aeronautical Decision-making and Single Pilot Resource
  Management. 316
  Training. 318
  Safety. 321
  Findings in the context of the theoretical framework. 323
Summary 326
CHAPTER V - Summary, Conclusions, and
Recommendations 329

Implications..332
    Implications based on the findings of RQ1.....................333
    Implications based on the findings of RQ2.....................338
    Implications based on the findings of RQ3 and RQ4. 346
    RQ1: Interpretation..352
    RQ2: Interpretation..353
    RQ3 and RQ4: Interpretation. 354
Recommendations..355
    Recommendations for practical applications...................356
    Future Research..357
Conclusions...359
Appendices **379**
    Appendix A: Data Analysis Procedures......................381
    Appendix B: Interview Question Guide.......................383
    Appendix C: Informed Consent Form.........................387
    Appendix D: Research Study and Participant Notification
   391
    Appendix E: Invitation Letter to the Panel of Experts .393
Index **395**
Curriculum Vitae...**407**
About The Author **431**
About The Book **433**

LIST OF FIGURES

Figure 1 – Japan Air Line (JAL), Boeing 747-446 Flight Deck; Photo courtesy of Altair78 18

Figure 2 – Airplane Cockpit; photo courtesy D. Toms 19

Figure 3 - Cockpit of the simulator for an Airbus A320 family aircraft; Photo Courtesy of Steve Jervetson......................27

Figure 4 - The cockpit panel of a Slingsby T67C light aircraft (UK registration G-BOCM) at Kemble Airfield, Gloucestershire, England. Photograph courtesy of Adrian Pingstone..........36

Figure 5 - Garmin G1000 panel of Diamond DA-42 Twin Star, N49494, based @ KBFI - Seattle, WA; Photo courtesy of Matthew Piatt..39

Figure 6 - Ice Air Glass Cockpit LCD Display; Photo courtesy of Monito1975 (Wikimedia.org). 46

Figure 7 - Primary Flight Display of a Boeing 747-400 aircraft. The electronic instrument shows airspeed, altitude, heading and additional data one a single display; photo courtesy of Markus Vitzethum..48

Figure 8 – Garmin 1000; Display Onboard SANSA Regional TI-BDZ. Above Golfo de Nicoya, Costa Rica; photo courtesy of Bernal Saborio..55

Figure 9 - An image of Honeywell's synthetic view for pilots; photo courtesy of the Honeywell Corporation. 93

Figure 10 - Current and prototype components of Coast Guard HC-130 Hercules legacy cockpit complete with upgraded flight displays, upgraded weather radar and yoke-mounted Garmin GPS MAP 695/696 portable aviation receivers at the Aviation Logistics Center in Elizabeth City, N.C., Oct. 21, 2010; photo courtesy of the U.S. Coast Guard and Lt. Cmdr. Brian Erickson. 98

Figure 11 – JAL Boeing 747-400, Flight Deck; Photo courtesy of Norio Nakayama, Saitama, Japan...................................111

Figure 12 - Six basic instruments in a light twin-engine airplane arranged in a "basic-T"; from top left: airspeed indicator, attitude indicator, altimeter, turn coordinator, heading indicator, and vertical speed indicator; photo courtesy of Meggar. 116

Figure 13 - Auto-pilot Panel; Photo courtesy of D-Laser. 125

Figure 14 - Instrument Training Flights; Photo courtesy of
Matthew Piatt..136
Figure 15 - Garmin G1000 panel of Diamond DA-42 Twin Star,
N49494, based @ KBFI - Seattle, WA; photo courtesy of
Matthew Piatt..144
Figure 16 - Cockpit of Northwest Airlines, DC-9-40 (reg. 9760).
Picture taken on the ground at Port Columbus International
Airport, Columbus, OH; photo courtesy of Dmitry
Denisenkov..164
Figure 17 - Cockpit of an Aeroflot Il-96-300; photo courtesy of
Dmitry Petrov..173
Figure 18 - The center cathode flight display in the cockpit of an
Airbus A320 family aircraft indicates spoiler deployment
and brake temperature as the aircraft vacates runway 22R
at NCE airport following landing; photo courtesy of Olivier
Cleynen. 184
Figure 19 - Primary Flight Display Garmin G1000; photo courtesy
of Caricato da Nubifer. 208
Figure 20 – Glass Cockpit; photo courtesy of Larre-Anthony. .222
Figure 21 – Professional Pilot in the USA; Photo courtesy of Ali
Rezaamidi. 245
Figure 22 - Operators in a control room pilot and monitor video
feeds from a remotely piloted UAV. Photo courtesy of
Gerald Nino, CBP, U.S. Dept. of Homeland Security – CBP.
....248
Figure 23 – Traffic Collision Avoidance System (TCAS) and
EHSI cockpit display; Photo courtesy of Jemr69.............255
Figure 24 - TCAS and IVSI cockpit display (monochrome); photo
courtesy of Mattes. 269
Figure 25 - Diagram showing the face of a true airspeed indicator
typical for a faster single engine aircraft; photo courtesy of
Mysid. 291
Figure 26 - Cockpit of a twinjet flight simulator; photo courtesy of
The DJ..304
Figure 27 - Upgraded "Glass" C-5M Instrument Panel; photo
courtesy of S. Voytek. 320
Figure 28 - The cockpit of the Concorde, which has an 'M'-
shaped yoke mounted on a control column; photo courtesy
of Christian Kath..341
Figure 29 - A Lufthansa Flight Training Bonanza 33 at Phoenix
Goodyear Airport USA; photo courtesy of Hinty.............351

LIST OF TABLES

Table 1 - Pilot Study Participants Data 241

Table 2 - Private Pilot's Demographic Data 245

Table 3 - Commercial- and Instrument-Rated Pilots Demographic
Data 247

Table 4 - Certified Flight Instructors Demographic Data 249

Table 5 - Comparison of Themes by the Embedded Unit of
Analysis 256

Table 6 - Compilation of Themes 257

Table 7 - Private Pilot Group Responses to Research Question 1
(RQ1) 260

Table 8 - Private Pilot Group Responses to Research Question 2
(RQ2) 265

Table 9 - Private Pilot Group Responses to Research Question 3
(RQ3) 267

Table 10 - Private Pilot Group Responses to Research Question
4 (RQ4) 267

Table 11 - Commercial-Instrument Pilot Group Responses to
Research Question 1 (RQ1) 270

Table 12 - Commercial-Instrument Pilot Group Responses to
Research Question 2 (RQ2) 273

Table 13 - Commercial-Instrument Pilot Group Responses to
Research Question 3 (RQ3) 278

Table 14 - Commercial-Instrument Pilot Group Responses to
Research Question 4 (RQ4) 281

Table 15 - Certified Flight Instructor Group Responses to
Research Question 1 (RQ1) 283

Table 16 - Certified Flight Instructor Group Responses to
Research Question 2 (RQ2) 285

Table 17 - Certified Flight Instructor Group Responses to
Research Question 3 (RQ3) 287

Table 18 - Certified Flight Instructor Group Responses to
Research Question 4 (RQ4) 289

Table 19 - Embedded Units of Analysis for Comparative Analysis
.... 292

ACKNOWLEDGEMENTS & FORWARD

I thank God for giving me the aptitude, devotion, humility, perseverance, and spiritual guidance to undergo such a daunting and humbling experience. I thank my family for understanding my absences at family gatherings because I was undertaking this enormous task. Your support during my frustration and tiredness were unequivocal and I am forever indebted to you.

This research is dedicated to dear a friend, Talmadge McClendon, Jr., who departed this world – too soon! The world is definitely a lesser place without his presence. Finally, yet importantly, I want to thank Dr. Robert George for chairing my Dissertation Committee, Dr. Martin Crossland, and Dr. Nicole Avena for providing scholastic expertise and assistance.

Calvin Nobles

ABSTRACT

Exploring Pilots' Experiences of Integrating Technologically Advanced Aircraft Within General Aviation: A Case Study

Calvin Nobles
Northcentral University, 2015

Dean of School: Dr. Peter Bemski, Ph.D.
Chair: Dr. Robert George, Ph.D.

A prevalent misconception is technologically
advanced aircraft (TAA) – those with so-called glass
cockpits – improved general aviation safety;
consequently, scientific research proved that TAA has
not improved aviation safety. A significant problem
with integrating TAA in general aviation was the lack
of a systematic approach for integrating TAA that
addressed the impact on current system designs,
processes, and regulations (Garibay & Young, 2013;
Pritchett, 2009; Robertson, 2010). This qualitative

single descriptive case study explored experiences of general aviation pilots on the integration of TAA in general aviation to determine how the automation of TAA can be used to increase aviation safety and improve pilots' decision-making skills.

Purposeful and snowball sampling were used to recruit 35 general aviation pilots from Georgia, South Carolina, and Florida. The participants were interviewed to gain insight on the integration of TAA, which resulted in the discovery of seven themes. The themes were: (a) training, (b) safety, (c) proficiency, competency, and familiarization, (d) transformative and challenging transition, (e) situational awareness, (f) decision-making, single pilot resource management and aeronautical decision-making, and (g) regulatory.

The themes, findings, and implications centered on the after-effects of integrating TAA, which were identified as adversely impacting aviation safety and pilots' decision-making abilities. The after-effects are: (a) outdated flight training modules, (b) dubious transition from conventional aircraft, (c) lack of regulatory oversight, and (d) failure to improve aviation safety and pilots' decision-making skills. The following recommendations are proposed to combat the after-effects of integrating TAA: (a) to continue strategic efforts amongst those in the aviation industry to improve flight training, (b) to leverage research to justify advanced flight training, (c) to evaluate TAA accidents to determine the technological impacts, and (d) encourage pilots to capitalize on free government-sponsored education and training.

Suggestions for future research include exploring impacts of Federal Aviation Administration (FAA) and Industry Training Standards (FITS), to examine the non-regulatory approach in general aviation, and to investigate a systematic process to implement technology that will have minimal aftereffects on pilot safety and decision-making.

Cirrus 22 Glass Cockpit; photo courtesy of Miltz311.

CHAPTER I

INTRODUCTION

In 2011, 92% of all aviation fatalities involved general aviation pilots (GAO, 2012; NTSB, 2010). A general aviation accident is as an erroneous flight occurrence by a general aviation pilot that results in extensive injuries to the aircraft occupants or substantial damage to the aircraft (GAO, 2012, Nilsson, 2011). The general aviation sector consists of flight training, search and rescue, aerial surveying and photography, crop dusting, recreational flying, and corporate aviation activities that are separate from military and commercial aviation operations (GAO, 2012; Sobieralski, 2013). A general aviation pilot receives a license by receiving 20 hours of dual flight instruction, attaining a minimum of 40 flight total hours, and passing an oral aeronautical examination and practical flight test (GAO, 2011). Of the 2,081 general aviation accidents that occurred between

2008 and 2010, 71% of the accidents resulted from multiple contributory factors, including pilot error, training deficiencies, and inputting erroneous flight controls (Craig, 2012; Franza & Fanjoy, 2012; GAO, 2012).

Figure 1 – Japan Air Line (JAL), Boeing 747-446 Flight Deck; Photo courtesy of Altair78

The integration of technologically advanced aircraft (TAA) in general aviation is changing general aviation (King, 2011). Whitehurst and Rantz (2012) defined TAA "as aircraft equipped with new-generation avionics that take full advantage of computing power and modern navigational aids to improve pilot awareness, system redundancy, and

depending upon equipment, improve flight deck

information about traffic, weather, and terrain" (p. 17).

Figure 2 – Airplane Cockpit; photo courtesy D. Toms

The integration of TAA is a necessity to

improve aviation safety, but the innovations are too

complex, which induced pilot confusion, and

automation and information overload that resulted in

accidents (FAA, 2013; Garibay & Young, 2013;

Robertson, 2010). To capitalize on the technological

advantages of TAA, pilots must undergo additional

training to learn to operate new-generation

innovations (Garibay & Young, 2013). Positive

aspects of TAA include: (a) advanced automated capabilities that allow pilots to fly further, (b) introduction to glass cockpits during the early stages of flight training, (c) cockpit instrumentation comparable to large jets, (d) automated in-flight weather, (e) autopilot capabilities, and (f) exposure to advanced flight training (GAO, 2012; Lindo et al., 2012; NTSB, 2010). The integration of TAA in general aviation has resulted in shortfalls in safety practices, flight training paradigms, regulatory, cockpit designs and equipment, and operational discrepancies, as well as government oversight concerns (NTSB, 2010) due to logistical, bureaucratic, and political obstacles (Garibay & Young, 2013).

A common technological deterministic practice is to resolve existing safety problems by integrating new technologies (Calvert, 2013; Chandler, 2012; NTSB, 2010). However, since the integration of TAA in general aviation, the number of fatalities in aviation accidents has more than doubled (McCracken, 2011). Aviation safety experts postulated that TAA have not improved aviation safety (NTSB, 2010). To improve general aviation safety, the Federal Aviation Administration (FAA) devised a strategic plan to

reduce general aviation accidents by 10% from 2009
to 2018 (GAO, 2012), which equates to an annual
reduction of 1% per year from 2009-2018, totaling a
10% reduction over ten years (GAO, 2012). To
achieve this goal, practitioners, researchers, general
aviation pilots, and aviation industry leaders need to
understand the impact on integrating TAA in general
aviation. Capturing the pilots' perceptions might lead
to an in-depth understanding of how technological
deterministic thinking applies to and affects general
aviation.

Background

In the United States, the general aviation
sector contributes 150 billion dollars to the national
economy and provides 1.26 million jobs (Cao & Ding,
2012). The general aviation sector is a thriving and
significant part of the aviation industry; therefore,
addressing the integration of TAA and glass cockpits
in general aviation is critical to the sustainment of
general aviation. For the purpose of this study, TAA
and glass cockpits[1] were used interchangeably. The

[1] *Aircraft with primary and multi-functional colored displays rather than
the conventional steam gauges and instruments, autopilot capable, a*

relevance of this study highlights the lack of
comprehension on how technology and automation
affects general aviation. Safety experts at the National
Transportation Safety Board (NTSB, 2010)
emphasized the integration of TAA in general aviation
has resulted in concerns with standardization of
equipment and design, flight-training deficiencies, and
improved accident investigation techniques. The FAA
and the general aviation community introduced FAA
and Industry Training Standards (FITS) as a training
platform to train general aviation pilots to operate TAA
(Halleran & Wiggins, 2010). FAA and Industry
Training Standards are flight training that incorporates
aeronautical decision-making, situational awareness
training, and risk management (Halleran & Wiggins,
2010). Even though FITS were designed in 2001,
general aviation is continuously plagued by accidents
(NTSB, 2010). On average, the NTSB investigates
about 1,600 general aviation accidents a year
(Garibay & Young, 2013). Garibay & Young (2013)
postulated that additional government oversight is
necessary because most general aviation accidents
occur in an unsupervised hierarchical construct and

*global positioning system with moving navigational maps, and an
advanced avionic suite (Mitchell et al, 2009).*

lack managerial checks and balances because most of general aviation flights are for personal and recreational use. Governmental oversight and industry involvement are vital to ensure general aviation pilots receive comprehensive training because the integration of TAA in general aviation has resulted in more fatalities (GAO, 2012; NTSB, 2010).

The FAA, as the leading regulatory authority in aviation implemented FITS to offset the training deficiencies in general aviation (Nilsson, 2011). The problems with FITS are stagnancy, general aviation pilots are not required to undergo subsequent training events, nor are there any regulations or policies mandating pilots to undergo a periodic review. Technologically advanced aircraft have the same advanced automated capabilities as commercial airliners or military jets; yet, there is no requirement mandating periodic proficiency training other than Code of Federal Regulations (CFR), Part 91 (Nilsson, 2011). The mandated requirements listed under Part 91 is basic and if not changed could enable the perpetuation of accidents in general aviation.

In a 2009 study on glass cockpits, researchers

determined the aviation community is uncertain of the
consequences from integrating TAA in general
aviation (Mitchell et al., 2009). The commercial pilots
that participated in the study complained of the
increasing complexities of automation, which affect
pilot proficiency, safety, workload, and human factors
(Mitchell et al., 2009). The general aviation sector
needs to capitalize on previously documented issues
by commercial pilots transitioning to aircraft with glass
cockpits in order to safeguard against persisting
problems from integrating TAA in general aviation
(Halleran & Wiggins, 2010). Commercial pilots
undergo more stringent training than general aviation
pilots do and commercial pilots work in crews to share
the workload; however, this is not the case for a
general aviation pilot who is the sole manipulator of
the aircraft.

In 2011, the FAA spent 203 million dollars on
general aviation, as an effort to improve safety by
modify training programs, and developing a
systematic approach to offset the adverse effects of
TAA (GAO, 2012; Halleran & Wiggins, 2010). Layton
(2012) postulated that aviation safety would decline
without innovative safety solutions because the

current approach to reducing aviation accidents is reactionary. The current approach is to prevent accidents by instituting measures to avert similar accidents (Layton, 2012). The reactive efforts are perpetuated by prioritization, redundant efforts, and the lack of communication (Layton, 2012). Based on Layton's supposition, the reactionary approach impacts the general aviation sector due to the lack of regulations and accountability. Recommending FITS to counter the problems associated with TAA is a reactionary approach because the training is not mandatory (Lindo et al., 2012). A common theme from the integration of TAA in general aviation is training because existing training curriculums lack the application for glass cockpits. Lindo et al. (2012) argued that the primary problem is pilots are not receiving adequate transition training, which is required when pilots upgrade to more advance and capable aircraft.

Another critical aspect with aviation safety is the dichotomy of the FAA's roles. Primarily, the FAA is charged with developing, implementing, and regulating aviation safety policies to safeguard against aviation safety accidents (Downer, 2012). The

FAA is responsible for influencing the economic verve of the aviation industry, which results in the continuous integration of new technologies in all sectors of the aviation industry (Downer, 2012). The dichotomous nature of counterbalancing these major responsibilities have negative influences such as the onset of new problems from technologies, outdated training paradigms, and the lack of an overarching strategic approach to implement technology in general aviation. Unless the FAA changes its organizational and cultural approach to regulating policy and managing innovations, it would be difficult to achieve both objectives simultaneously.

Figure 3 - Cockpit of the simulator for an Airbus A320 family aircraft; Photo Courtesy of Steve Jervetson.

Statement of the Problem

A significant problem with integrating TAA in general aviation is the lack of a systematic approach for integrating TAA that addresses impacts on current system designs, processes, and regulations (Garibay & Young, 2013; Pritchett, 2009; Robertson, 2010). The general aviation sector needs a systematic process to integrate technologies without introducing new hazards and safety concerns (Di Renzo, 2010). Without a systematic process, the general aviation sector lacks the ability to identify hazards and implement corrective measures to eliminate unintended consequences such as aviation accident fatalities, improper training, and the lack of regulatory oversight (Cassens, Young, & Greenan, 2011; McCracken, 2011; NTSB, 2012). Robertson's study determined the automated capabilities needed for general aviation aircraft as an effort to target pilot error because 80% of general aviation accidents were due to pilot error. Robertson recommended additional research on general aviation pilots' trust and rational utilization of automation. In another study, McCracken's (2011) researched general aviation pilots transitioning from conventional aircraft to TAA, which majority of the pilots received substandard scores. McCracken's study substantiated Robertson's research that general aviation pilots were uncomfortable operating TAA after completing initial orientation training. McCracken recommended research on pilots transitioning to TAA to identify changes to improve the quality of flight training. Garibay's and Young's (2013) study highlighted the importance of using technologies to improve aviation safety by comparing general aviation and commercial aviation operational procedures. Garibay and Young postulated that TAA improve situational awareness and general aviation pilots need to embrace new technologies. Garibay and Young recommended follow-on research on the efficacy of recurring flight training, improved safety initiatives, and the use of technology to reduce aviation accidents in general aviation. These studies highlighted increased accident fatalities, improper training, and inadequate regulatory oversight due to the lack of a systematic approach to highlight the impacts.

Purpose of the Study

The purpose of this qualitative single
descriptive case study was to explore the experiences
of general aviation pilots on the integration of TAA
and to determine how the automation of TAA can be
used to increase aviation safety and improve pilot
decision-making skills. A qualitative single descriptive
case study was conducted using semi-structured
interviews, technical reports, notes from the
interviews, data from flight logbooks to explore pilots'
experiences on the integration of TAA and how to use
TAA to improve aviation safety and pilots' decision-
making skills. In order to gain comprehensive
perspectives, this case study explored a single case
with three embedded units of analysis. This case
study explored a single case with three embedded
units of analysis, which were three distinct categories
of general aviation pilots. The first unit of analysis was
private pilots, the second unit of analysis was
instrument and commercial-rated pilots, and the third
unit of analysis was certified flight instructors. The
embedded units of analysis or participant groups
provided data on how the integration of TAA impacted

private pilots, instrument and commercial rated pilots, and certified flight instructors (Rao, 2013). In this study, embedded units of analysis and multiple sources of data were used to help capture the impact of integrating TAA in general aviation. Using embedded units of analysis and multiple sources of data in a single case study increased the trustworthiness of the study (Rao, 2013, Yin, 2009). Face-to-face interviews are a typical form of data collection in case studies (Baxter & Jack, 2008; Yin, 2009). In this study, general aviation pilots in Florida, Georgia, and South Carolina were interviewed face-to-face to explore the integration of TAA in general aviation. The questions for the semi-structured interviews were specifically designed to explore the integration of TAA in general aviation. Purposeful and snowball sampling strategies were used to select the participants for the research study. In qualitative research, research participant numbers vary, so the targeted purposeful sample was 20 to 30 pilots from a population of 77,842 general aviation pilots in Florida, Georgia, and South Carolina ("FAA Certificated Pilots", 2011). Some qualitative researchers deemed 10 or fewer research participants are sufficient to reach data saturation (Cane, McCarthy, & Halawi,

2010; Creswell, 2009). Data saturation occurs when

data collection no longer reveals any new themes,

perspectives, insights, or information (Suri, 2011).

The researcher interviewed 35 research participants

before reaching data saturation. Notifications of the

study were distributed via Internet and telephonic

channels, and dissemination of study

announcements. The notifications included details

about the study and the applicable contact information

for individuals interested in participating in the study.

During the initial phone call or e-mail, the primary

objective was to inquire about the potential research

participant's experience with glass cockpits to make

sure the pilot has adequate TAA experience. To

participate in this study, the research participants had

a private pilot certificate (minimum), and at least 10

flight hours in TAA. Having 10 flight hours ensured the

pilots had the minimum knowledge to manipulate the

aircraft. Therefore, it was essential to verify and

review the pilots' qualifications and flight logbooks.

The goal was for the general aviation pilots to have

various experience levels ranging from private pilots

to certified flight instructors to provide in-depth

descriptions on the integration of TAA in general aviation. The interview was a 60-minute, face-to-face session in a semi-structured format to ask each research participant open-ended questions, which was digitally video-recorded after getting permission. Selecting research participants in Florida, Georgia, and South Carolina made it logistically possible to travel and conduct face-to-face interviews. The interview data were corroborated with the pilot's flight logbook data, technical reports, and observation notes. Specifically, the technical reports were the: (a) Aircraft Owners and Pilot Association's Air Safety Foundation (ASF) TAA Safety and Training Reports (2005 and 2007), (b) the ASF Nall Reports from 2012–2013, the NTSB Report on the Introduction of Glass Cockpits (2010), (c) the Government Accountability Office (GAO) Report on General Aviation Safety (2012), (d) the GAO Report on Initial Pilot Training (2011), and (e) a 2013 study on reducing general aviation accidents by utilizing airline operational strategies. The collected data was analyzed for major themes and underwent a case analysis comparison between the participants groups to understand the impact of TAA on general aviation and to determine how to use the automation of TAA to improve aviation safety and pilots' decision-making skills. The case study design was an effective method to attain contextual reflections (Miller, 2011; Yin, 2009) on the impact of integrating TAA in general aviation. The case study design allowed exploration of multiple units of analysis to increase robustness of the study and reduced the pre-dispositions of a using a singular unit of analysis (Rao, 2013; Yin, 2009). This qualitative single embedded descriptive case study explored the integration of TAA in general aviation and to determine how to use the automation of TAA to improve aviation safety and pilots' decision-making skills, which led to recommendations on improving the integration of TAA in GA and pilots' decision-making skills.

Theoretical Framework

The theoretical framework for this study was based primarily on technological determinism. The basis of technological determinism is the creation and implementation of new technologies are intended to improve human lives by enhancing processes and procedures (Chandler, 2012), regardless of the cultural and societal affects (Yang, 2009). Exploring technological deterministic thinking and its impact from a micro-level is an unusual undertaking because technology and social science scholars typically evaluated technological determinism from a macro-level while disregarding individual-level influences and micro-level transformation (Lane & Lyle, 2011). Technological determinism is reductionist in nature in which technology acts as the agent that influences changes (Jackson & Philip, 2010). Jackson and Philip (2010) criticized technological determinists for overlooking the significance of organizational culture. Researchers primarily viewed technological determinism from a macro-level to determine social and cultural transformations that are determined by technological developments (Chandler, 2012). Technological determinism is not without limitations; scholarly research is necessary to account for the social perspectives such as social needs, economic interests, political implications, and decision-making impacts (Selwyn, 2012). Understanding the social perspectives enabled social science researchers to advance technological deterministic thinking to account for impacts at micro-levels (Selwyn, 2012).

Technological deterministic thinking supports the philosophical view that implementing technologies causes changes, in which consumers have little to almost no control (Chandler, 2012). Technological determinists believe the development of technology is an autonomous activity that drives the continuous integration of technological innovations that may shape and improve society, (Douglas, 2010; Oliver 2011; Yang 2009), even though some aviation innovations have unintended consequences (Doorn, 2012; Parasuraman & Manzey, 2010).

Historical evidence indicated industrialization negatively affected cultures because communities once produced technologies based on cultural values and necessities (Vanderburg, 2012). Industrialization transitioned efforts away from community-based technologies to universal-based technologies, which disregards cultural values and needs (Vanderburg, 2012). Technological deterministic thinking supports universality, which increased the incompatibility of technologies in some disciplines (Vanderburg, 2012) while some innovations introduced unintended consequences (Doorn, 2012).

*Figure 4 - The cockpit panel of a Slingsby T67C light aircraft (UK
registration G-BOCM) at Kemble Airfield, Gloucestershire,
England. Photograph courtesy of Adrian Pingstone.*

Understanding the foundational underpinnings
of technological determinism is essential to knowing
the implications to society, organizations, and
institutions (Vanderburg, 2012). Technological
determinism thinking is applicable to business
practices, socialization, culture and anthropology,
industrialization, science, politics, and the military
(Vanderburg, 2012). Technology scholars believe that
"technology is changing people" and in some cases
innovations have alienated humans (Vanderburg,

2012, p. 27). Chandler (2012) expounded that some people use technology to gain the advantage from an educational, social, or professional perspective. In aviation, technology enables efforts to improve safety (Lindo et al., 2012). Some researchers argue normalization of technology occurs when innovations are commonly used; therefore, refraining from utilizing the specified technology creates a contrariety (Chandler, 2012).

Hallström and Gyberg (2011) postulated a technology model consists of the internal and external actors, political factors, economic drivers, social actors, laws, and regulations, cultures, and scientific and technological requirements. Even though Hallström and Gyberg developed a model for justifying technology, Chandler (2012) argued the development of technology is an autonomous force. Chandler postulated that interfering with technological developments disturb the natural flow of innovations. The primary driver of technological developments is necessity; however, in some cases, innovations like the telephone created a need (Hallström & Gyberg, 2011). Researchers stressed the failure of design coupled with human dissatisfaction is the driving force behind innovations rather than the application of science (Hallström & Gyberg, 2011). For example, in a study conducted by Garibay and Young (2013), the researchers recommended improved technology to help reduce the number of general aviation accidents is what Chandler (2012) characterized as "limited powers of foresight and understanding, familiar problems of political governance and coordination, and from a fundamental ideological embrace of technological progress" (p. 262).

Another theory akin to technological determinism is technological imperative. Webster (2013) stated that technological imperative pertains to fundamental postulations that technology has a dominating persuasion, which is inescapable and overpowering. The nature of technological imperative is to coerce adoption of emerging technological developments or users become irrelevant for refusing new technologies (Webster, 2013). The technological deterministic rhetoric that supports technological imperative subdues opposing ideology that downplays the significance of emerging technologies (Webster, 2013).

Figure 5 - Garmin G1000 panel of Diamond DA-42 Twin Star, N49494, based @ KBFI - Seattle, WA; Photo courtesy of Matthew Piatt.

According to Webster (2013), the basis of technological imperative is "if a technology can be developed it ought to be developed, and will be developed, without regard for ethical considerations or making value judgments about the technology" (p. 95). This non-empathetic approach is visibly evident in the aviation industry. For example, the aviation infrastructure is a system of systems, comprised of tightly integrated components that new technologies often disrupt, which disregard institutional practices, the social tenets, and policies (Batteau, 2010; Craig, 2012; Ropp, 2009). The automated aspects of new technologies are causing human factors

consternations (Parasuraman & Manzey, 2010).

Human factors pertain to the science and application

of human performance in an operational environment

consisting of methods and procedures to evaluate

factors that adversely affect performance (Bowen,

Sabin, & Patankar, 2011). Some researchers argued

the integration of TAA in general aviation has affected

flight-training paradigms, creating regulatory and

government oversight concerns, and human factors

problems due to the constant integration of new

technologies (Lindo et al., 2012). The aforementioned

impacts on general aviation are indicative of

technological deterministic thinking (Chandler, 2012).

In general aviation, these unintended consequences

come in the form of more in-depth training,

regulations, oversight, and an improved

understanding of human-machine interaction (Lindo et

al., 2012; Parasuraman & Manzey, 2010).

Another theory analogous to technological

determinism is the theory of technology cycles, which

pertains to the social interaction between

organizations, communities, and individual actors that

chart the course for technological developments

(Dokko, Nigam, & Rosenkopf, 2012). A key element of the technology cycles theory is the evolvement evolution of institutional, organizational, and individual actors; discourse between technological actors shaped the path for future technological developments (Dokko, Nigam, & Rosenkopf, 2012). The negative interaction amongst technological and community actors often disrupted the dominant design or construct of technology, which is followed by discontinuity and incremental changes to technological developments (Dokko, Nigam, & Rosenkopf, 2012). Dominant designs were the outcome of extensive collaboration and discourse between technological actors and communities with political and social interests that drove changes to current technologies (Dokko, Nigam, & Rosenkopf, 2012).

Technological deterministic thinking perpetuated the trends in general aviation because the perceptions on integrating TAA in general aviation were not explored. Technological determinism influenced socio-technical systems. Erichsen et al. (2013) asserted the term socio-technical systems was coined in the 1950s to address the intricacy of integrated systems that involved humans, machines, and the organized activities within the complex systems. Socio-technical systems focused primarily on humans, the organization, and the technological features used to accomplish system objectives (Erichsen et al., 2013). The national airspace system is a socio-technical system that requires expertise, operator training, regulatory oversight, and precise interaction between pilots, air traffic controllers, ground controllers, and aviation specialists to achieve a high degree of functionality and precision (Erichsen et al., 2013; Sharma, Coit, Oztekin, & Luxor, 2009). The integration of new technologies affected the operability and functionality of socio-technical systems because humans had to undergo significant training and engineers had to dissect the socio-technical system to determine the impact on each component of the system (Strauch, 2010). Strauch emphasized new technologies affect cultures by changing commonalities and behaviors of people who operate socio-technical systems. Strauch asserted, "aviation operations comprise many segments, such as military, recreational, agricultural, and commercial, which differ in complexity, type and extent of operator training and experience, operating procedures, and company or government oversight, among other factors" (p. 253). Researchers and engineers have to scrutinize the implication of technological developments to evaluate the impacts on the existing aviation infrastructure. Modifying aviation socio-technical systems influenced the communication, coordination, and decision-making of the system operators (Strauch, 2013).

Technological deterministic thinking affected socio-technical systems through the integration of technological developments; thus, making this evolution continuous (Chandler, 2012; Dokko, Nigam, & Rosenkopf, 2012; Strauch, 2013). Technological deterministic thinking were interrelated through the continuous integration of technologies and the persistent discourse between technological communities, actors, and individuals that advocated for technological changes, which affected socio-technical systems (Chandler, 2012; Dokko, Nigam, & Rosenkopf, 2012; Strauch, 2013). From a socio-technical system approach, new technologies alter the culture, cognitive, operational, and social factors of complex systems because innovations change the interoperability, training requirements, reliance, and human factors of the system (Chandler, 2012; Straus, 2012).

Another concern and impact of technological deterministic views is researchers reprobated the theory for its myopic view and disregard for individual-level effect on micro-level transformation and the adoption of innovations (Lane & Lyle, 2011). The narrowed-minded approach on integrating automated capabilities in general aviation was obvious by the FAA implementing FAA and Industry Training Standard (FITS), as a training methodology to offset the problems caused by the integrating technologies in general aviation (Halleran & Wiggins, 2010). Studies on TAA revealed the increased complication of automation resulted in more flight anomalies; thus, requiring pilots to undergo additional training to manage the advanced automated capabilities (Halleran & Wiggins, 2010).

Technological deterministic thinking is a dominant theory that perpetuates the development of technologies (Vogel, 2013) regardless of the negative consequences (Chandler, 2012). Technological determinism thinking was an acceptable philosophy amongst policy makers due to its straightforward underpinning (Vogel, 2013). The continuous integration of technologies in general aviation was an effort to improve aviation safety (Oster, Strong, & Zorn, 2013). Human-machine interaction, atrophy of basic flying skills, lack of cockpit standardization, resource management issues, and concerns of technology failures were problems associated with new aviation technologies (Oster, Strong, & Zorn, 2013). The problem with technological deterministic thinking was the disregard to implement measures, processes, or predictive analysis to reduce problems associated with emerging technologies (Chandler, 2012). For example, safety experts postulated TAA did not improve aviation safety (NTSB, 2012). This approach disregarded cultural and organizational practices, consequently, leading some researchers to dispute the effectiveness of new technologies on reducing general aviation accidents (GAO, 2012). This was significant because 90% of new general aviation aircraft have glass cockpits (NTSB, 2010). Technological developments were negatively impacting flight training, human factors, cognitive development, cockpit designs, and strategic safety

initiatives of general aviation due to the rapid and
constant implementation of new technologies (Lindo
et al., 2012; Novicevic, Hayek, Buckley, &
Humphreys, 2009). Even with the integration of TAA,
the negative consequences cannot be ignored
because 1,600 general aviation accidents are
investigated annually (Garibay & Young, 2013; NTSB,
2012).

*Figure 6 - Ice Air Glass Cockpit LCD Display; Photo courtesy of
Monito1975 (Wikimedia.org).*

The foundational theories for integrating TAA in
general aviation were inadequate as evident by
members of the NTSB investigating over 1,600

general aviation accidents per year because the goal of integrating technology was to improve aviation safety (Garibay & Young, 2013). Technological determinism thinking and technology cycles were misapplied in the process of integrating TAA in general aviation because technologists disregarded the impacts on training, current processes, and disruptions to aviation systems (Chandler, 2012, Craig, 2012; Dokko, Nigam, & Rosenkopf, 2012). The study extended technology determinism theoretical application to the integration of TAA in general aviation. Technological determinism was the belief that innovations serve to improve and advance processes, industries, or human-machine interaction (Chandler, 2012). The integration of TAA in general aviation has not improved aviation safety, according safety experts the number of fatalities involving TAA has increased (Groff, 2010a; Groff, 2010b). Some researchers believe TAA has not added any value to general aviation safety (Groff, 2010b). Technological deterministic thinking has resulted in a higher fatality rate due to the integration of TAA in general aviation (Groff, 2010b), which was indicative the existing theoretical beliefs and practices were flawed. The integration of TAA in general aviation contradicted technological deterministic thinking; therefore, this study might result in changes to the theoretical and philosophical beliefs of technologists, researchers, government officials, and aviation leaders.

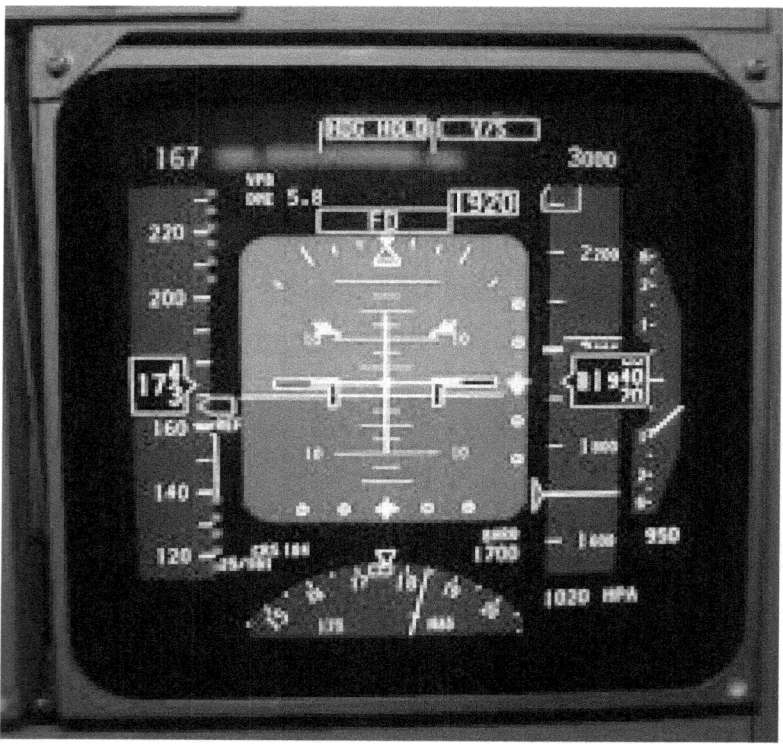

Figure 7 - Primary Flight Display of a Boeing 747-400 aircraft. The electronic instrument shows airspeed, altitude, heading and additional data one a single display; photo courtesy of Markus Vitzethum.

Research Questions

The purpose of this qualitative single descriptive case study was to explore the experiences of general aviation pilots on the integration of TAA and to determine how the automation of TAA can be used to increase aviation safety and improve pilot decision-making skills. The objective of qualitative research questions was to explore the phenomenon first-hand by interviewing general aviation pilots to develop an in-depth understanding of integrating TAA in general aviation (Creswell, 2009; Vasconcelos, 2010), which this data along with technical reports and studies, and observation notes were the collected data for the study (Yin, 2009). From the collected data, contextual information in the form of themes, patterns, and categories were extracted to develop the case (Yin, 2009). The intent of the qualitative research questions was to provide the best means to gather data for the case study. Each research question served to gather enriched data on the participants' first-hand experiences with the impact of integrating TAA in general aviation. The following

central question and sub-questions explored the
integration of TAA in general aviation:

Q1. What are the experiences of general
aviation pilots regarding the integration of TAA?

Q2. How can general aviation pilots use the
automation of TAA to reduce accidents?

Q3. How can general aviation pilots reduce the
effects of technological innovations to capitalize on
the positive attributes of TAA?

Q4. What is the impact of TAA on aeronautical
decision-making skills of general aviation pilots?

Nature of the Study

The purpose of this qualitative single
descriptive case study was to explore the experiences
of general aviation pilots on the integration of TAA
and to determine how the automation of TAA can be
used to increase aviation safety and improve pilot
decision-making skills. Another reason to pursue this
study was because "technology's impact on societies
and individuals highlights an area that is presently
understudied; this need is amplified by reviews of the

field and calls for future research" (Cilesiz, 2011, p. 489). This qualitative case study methodology replicated logic to ensure the units of analysis forecasted analogous results (Yin, 2009). An advantage of using a qualitative, single-descriptive, case study was an opportunity to interview and collect data on general aviation pilots to explore the impact of integrating TAA in general aviation (Baxter & Jack, 2008; Yin, 2009).

Purposeful and snowball sampling techniques were used to target general aviation pilots. Purposeful sampling involves using individuals in the field for the intent of identifying information-rich cases associated with the phenomenon (Cane, McCarthy, & Halawi, 2010; Creswell, 2009; Suri, 2009). Purposeful sampling guided efforts to target a sample population to include general aviation pilots to ensure participants have lived experiences in order to contribute to understanding the phenomenon (Leedy & Ormrod, 2005). The purposeful sample involved 35 research participants from a population of 77,842 general aviation pilots in Florida, Georgia, and South Carolina ("FAA Certificated Pilots", 2011).

Snowball sampling in conjunction with purposeful sampling ensured the most credible and appropriate participants are targeted (Baltar & Brunet, 2012). Snowball sampling is a method in which one research participant provides the name and credentials of another potential research participant that meets the sample criteria, which happens continuously (Baltar & Brunet, 2012). Baltar and Brunet postulated snowball sampling is an effective methodology used in qualitative and descriptive studies in which confidence and trust are necessary to locate specific participants for a study. To overcome geographical and time barriers associated with snowball sampling (Baltar & Brunet, 2012) this study was limited to 77,842 general aviation pilots in Georgia, Florida, and South Carolina ("FAA Certificated Pilots"). Snowball sampling provided the basis for targeting general aviation pilots to explore the integration of TAA. Official notifications of the study were disseminated to flight schools, flying clubs, flight business offices, and general aviation pilots with a recommendation to promulgate to any general aviation pilot in Georgia, Florida, or South Carolina with TAA experience.

Data collection included semi-structured interviews, technical reports, flight logbooks, and observation notes during the interviews. The semi-structured interviews allowed the participants to answer open-ended questions, which allowed the restructuring of previously asked questions, probing, asking follow-up questions, and interjecting with additional questions for further clarification (Kvale & Brinkman, 2009; Yin, 2009). Each participant underwent a 60-minute, face-to-face interview to provide invaluable details on flight experiences and affiliation with general aviation. Face-to-face interviews allowed the direct observation and documentation of body language, facial expressions, and other forms of non-verbal activity (Kvale & Brinkman, 2009). The researcher asked the participants open-ended questions to gain insight on the impact of integrating TAA in general aviation and to determine how to leverage the automation of TAA to improve aviation safety and pilot decision-making skills. The qualitative research questions provided the best means to gather data for the described study. Each research question served to gather rich data on the research participants' first-hand experiences with the phenomenon.

In a qualitative case study, the goal is to achieve data saturation, which will determine the number of research participants (Hathorn, Machtmes, & Tillman, 2009). Data saturation occurs when data collection no longer reveals any new themes, perspectives, insights, or information (Suri, 2011). Private pilots, instrument and commercial rated pilots, and certified flight instructors were essential to this study because as the embedded units of analysis and research participants, each pilot provided a personal account on the impact of integrating TAA in general aviation during a semi-structured interview. The semi-structured interviews were corroborated with technical reports and studies, flight logbooks and observation notes to highlight themes that were associated with the phenomenon. The collected data was analyzed using a modified van Kaam approach.

Using a modified van Kaam approach to analyze the data appropriate as a method to accomplish the analysis (Moustakas, 1994) see Appendix A. The modified van Kaam model included the following steps: (a) electronically record the interviews, (b) transcribing the interviews, (c) member checking transcripts, (d) coding the transcripts,

bracketing, and reduction, (e) repeating the coding

process and dividing the data into phrases and

categories, (f) identifying themes, (g) developing

textual and structural descriptions, (h) writing

summary of each interview, (i) conducting second

interview (if necessary), (j) conceptualizing and

consolidating themes, (k) writing a composite

summary, and (l) reviewing audit trail (Hycner, 1985;

Moustakas, 1994; Tirgari, 2012).

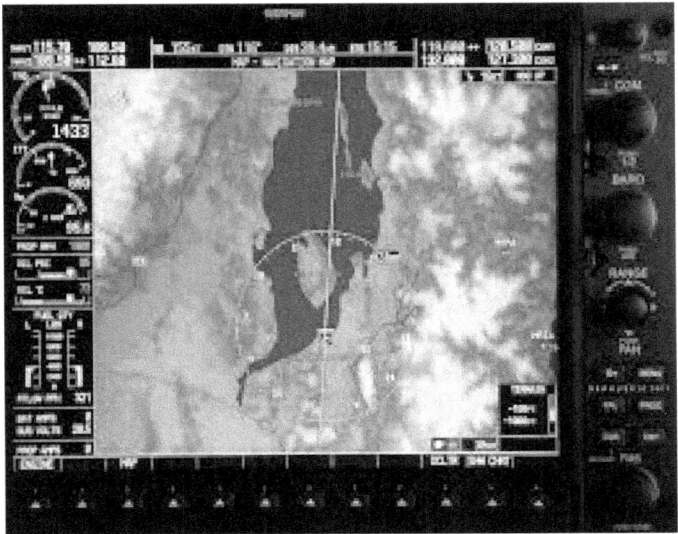

*Figure 8 – Garmin 1000; Display Onboard SANSA Regional TI-
BDZ. Above Golfo de Nicoya, Costa Rica; photo courtesy of
Bernal Saborio.*

Significance of the Study

Technological deterministic thinking is a rigid theoretical belief that disregards negative implications (Chandler, 2012) such as unintended outcomes like accidents, fatalities, a critical training deficiency, and a lack of government oversight (Cassens, Young, & Greenan, 2011; McCracken, 2011, NTSB, 2012). Some technological determinists believed impeding the integration of innovation disrupted the natural progression of society (Chandler, 2012). Failure to assess the cultural, social, and institutional consequences associated with new technologies perpetuates the theoretical misunderstanding of technological determinism. The misunderstanding of technological deterministic thinking resulted in a skewed theoretical comprehension that resulted in negative consequences from integrating TAA in general aviation. The integration of automated technologies in general aviation has resulted in unintended consequences such as an increase in accident fatalities, improper training, and the lack of regulatory oversight, which negatively affect safety in general aviation (Cassens, Young, & Greenan, 2011; McCracken, 2011; NTSB, 2012). The general aviation community is continuously plagued by aviation accidents, compounded by integration of TAA in the general aviation sector. A recent 2013 report by aviation experts indicated technologies caused automation dependency, eroded basic flying skills, confused pilots during emergencies, and illustrated a lack of proficiency in using automated capabilities (FAA, 2013). Seventy percent of general aviation accidents were due to pilot error (Pourdehnad & Smith, 2012). Aviation experts, scholars, and industry leaders are working to develop initiatives to reduce human induced accidents (King, 2011). The integration of TAA allows general aviation pilots to operate aircraft (12,500 pounds or less) with the same advanced capabilities found in commercial jets (Halleran & Wiggins, 2010). A major concern was human factors, because scientific research proved automation reduces human's system knowledge, generates mistrust, causes overreliance, and degrades contentment (Pritchett, 2009). Overreliance, mistrust, and diminishing system knowledge were plausible causal factors in accidents involving TAA.

The purpose of this qualitative single descriptive case study was to explore the integration of TAA in general aviation.

As general aviation aircraft become more complex, the higher order of thinking skills for general aviation pilots is necessary to comprehend the level of automation required to fly TAA safely. Two primary causes of general aviation accidents were the lack situational awareness and poor aeronautical decision-making (Nilsson, 2011). This was indicative of outdated flight training programs, poor government oversight, and ineffective management process (King, 2011; Layton, 2012) because aviation technologies continue to outpace the processes to vet innovations for general aviation. There were ongoing initiatives to improve training like FITS and single-pilot resource management; however, the regulations do not mandate for pilots that fly TAA undergo proficiency training on a routine basis as commercial and military pilots do (Hendrickson, 2009).

According to Franza and Fanjoy (2012), aircraft with glass cockpits have not improved safety because some general aviation pilots were not properly trained to operate TAA. To achieve the appropriate level of

training and regulatory changes needed to govern
technologies in general aviation require an
institutional change to organizational learning and a
holistic process to implement training and regulations
requirements due to the changes occurring in general
aviation (Franza & Fanjoy, 2012; Garibay & Young,
2013; McCracken, 2011). This study might change
the theoretical underpinnings and philosophical
beliefs associated with implementing TAA in general
aviation. The current theoretical perspective included
shortfalls as evident by the problems associated with
integrating TAA in general aviation. Inadequacies of
mandated training, lack of oversight, the disjointed
efforts between general aviation organizations, and
increasing fatalities and accidents in general aviation
requires new theoretical perspectives to reduce the
negative consequences associated with implementing
TAA in general aviation.

Definition of Key Terms

Aeronautical decision-making. Aeronautical
decision-making is a methodical approach for
mentally processing the most appropriate course of

action based on a specified situation that the pilot is experiencing (Cassens et al., 2011).

Automation management. Automation management is the pilot's ability to navigate and control the aircraft via the aircraft automated systems (Pritchett, 2009). Pilots have to understand when to leverage the automated capabilities and when to manipulate the aircraft to prevent the pilot from using the automation incorrectly (Pritchett, 2009).

Aviation accident. An aviation accident is an incident involving the operation of an aircraft intended for flight in which a death or a grave injury occurs, or the aircraft receives substantial damage (NTSB, 2010).

Aviation Safety Program. The Aviation Safety Program is the partnership of the National Aeronautics and Space Administration and the FAA, as authorized by the White House Commission on Aviation Safety and Security to collaborate and pursue ways to improve aviation research and technologies to enhance aviation safety within the

National Airspace System (Sharma, Coit, Oztekin, &
Luxor, 2009).

Conventional aircraft. Conventional aircraft are
airplanes with analog gauges; therefore, lacking
automation, avionics, and global position system with
navigation maps (Lindo et al., 2012). These aircraft
are interchangeable with non-TAA (Lindo et al.,
2012).

Crew resource management. Crew resource
management (CRM) is a team concept to operate the
aircraft effectively, which includes single pilots flying
small aircraft, using all available resources through
each phase of flight by making all crewmembers
decision makers and contributors to the safe
operation of the aircraft (FAA, 2008). Pilots,
dispatchers, cabin crewmembers, maintenance
personnel, and air traffic controllers are also critical
members of the CRM process (Halleran & Wiggins,
2010).

Fixed-wing aircraft. Fixed-wing aircraft are
aircraft with wings secured to the aircraft structure,

which excludes rotorcraft; therefore, for the purpose
of this study, it pertains to aircraft weighing less than
12,500 pounds (Nilsson, 2011).

General aviation. The United States
Government Accountability Office considers general
aviation as civil aviation, which include all
components of aviation excluding air transport
carriers operating under a certification to provide air
transportation to the public with authorization from the
Civil Aeronautics Board (GAO, 2011). General
aviation is the largest sector of all aviation sectors in
regards to airports, aircraft, and pilots, which includes
over 91% of all aircraft, 75% of operations at FAA-
towered and non-towered airports, and 80% of
certified pilots in the United States are general
aviation pilots (GAO, 2011). General aviation includes
corporate, business, personal, flight instruction, aerial
application, aerial observation, sightseeing, touring,
air taxi, medical, weather modification flights, sale
demonstrations, flight-testing, and research and
development flights (GAO, 2011).

Glass cockpit. Glass cockpit refers to aircraft with primary and multi-functional colored displays rather than the conventional steam gauges and instruments, autopilot capable, a global positioning system with moving navigational maps, and an advanced avionic suite (Mitchell et al, 2009). For the purpose of this research, glass cockpit equipped aircraft and TAA are interchangeable. Additionally, glass cockpits provide pilots with computerized data such as navigational data, digital engine control data, global positioning systems with moving maps, in-flight weather, and fuel management data (Mitchell et al., 2009; NTSB, 2010).

Higher order thinking skills. Higher order thinking skills are the cognitive process and mental aptitude involved in pilots making logical decisions on specific situations (Dimitriadis, 2012). Some social scientists and researchers believe there is no universal definition of higher order thinking skills (Robertson, 2010). Therefore, for the purpose of this study, higher order thinking skills pertained to pilots possessing the cognitive and critical thinking capacity to aviate and navigate the aircraft, as well as the ability to understand the flight characteristics of the aircraft and its automated capabilities (Dimitriadis, 2012; Halleran & Wiggins, 2010).

Human factors. Human factors pertain to the science and application of human performance in an operational environment consisting of methods and procedures to evaluate factors that adversely affect performance (Bowen, Sabin, & Patankar, 2011).

National Airspace System. The National Airspace System is the collective network of the U.S. airspace, air navigation facilities, airline operation centers, equipment specialists, dispatchers, meteorologists, weather centers, pilots, air traffic controllers, aviation services, regulations, aviation information, personnel, and materials that support air operations in the United States (Berry and Pace, 2011).

National Transportation Safety Board (NTSB). The NTSB is an autonomous organization that is chartered by the United States Congress to investigate aviation, marine, highway, and railroad accidents (Lewis, 2014).

Pilot error. Pilot error is considered erroneous decision making that result in an aviation accident or incident (Halleran & Wiggins, 2010).

Risk management. Risk management is the decision-making process that a pilot uses to recognize hazards, evaluate the situation, and determine the best course of action (Cassens et al., 2011).

Safety regulations. Safety regulations are Federal Aviation Regulations that provide specific guidance to enhance safety or reduce ambiguous understanding of safety regulations (NTSB, 2010). For the purpose of this study, safety regulations pertain to provisions or guidelines listed in the Federal Aviation Regulations.

Scenario-based training. Scenario-based training is a training methodology that incorporates real world scenarios to train pilots in an operational environment to deal with emergencies in during flight operations (Cassens et al., 2011).

Single-pilot resource management. Single-pilot resource management is the ability to manage internal and external resources available to a single pilot operating an aircraft to ensure the safe flying and operation of the aircraft (Cassens et al., 2011).

Situational Awareness. Situational awareness is the accurate assessment and comprehension of the available information and circumstances in the fundamental risk areas that influence safety during flight operation (Cassens et al., 2011).

Technological determinism. Technological determinism is the concept that technology influences society and culture and humans have no control on technological influence (Chandler, 2012). Technological determinists believe that the development of technology is an autonomous force that drives the continuous integration of technological innovations to shapes society and improves lifestyles (Yang, 2009).

Summary

The purpose of this qualitative single descriptive case study was to explore the experiences of general aviation pilots on the integration of TAA and to determine how the automation of TAA can be used to increase aviation safety and improve pilot decision-making skills. General aviation pilots in Florida, Georgia, and South Carolina were interviewed to gain first-hand knowledge on the impact of TAA in general aviation. Integrating TAA in general aviation resulted in unintended outcomes like accidents, fatalities, a critical training deficiency, and a lack of government oversight (Cassens, Young, & Greenan, 2011; McCracken, 2011, NTSB, 2012). The emergence of TAA in general aviation exposed flight-training issues, safety concerns, human factors problems, and the lack of regulations and governmental oversight (Nilsson, 2011). A finding by the NTSB (2010) revealed TAA did not improve safety in general aviation.

The general aviation sector struggled to capitalize on the advanced automation of TAA

because of antiquated training practices and
government stakeholders' inability to keep pace with
implementation of new technologies. Seventy percent
of general aviation accidents were due to pilot error
and with the ascendancy of aircraft with glass
cockpits, the general aviation sector needs to undergo
an institutional and cultural change to leverage
initiatives like FITS, single pilot resource
management, and aeronautical decision-making
(Nilsson, 2009; NTSB, 2010). Technological
deterministic thinking and the disregard for the
negative consequences was the result of misleading
theoretical perspectives that drive the constant
implementation of new technologies. Technological
deterministic thinking would continue to cause
mayhem unless a new theoretical understanding is
developed to determine the best technological
implementation process for general aviation.

Currently, the process is reactive and lacks a
systematic approach and pre-phasing testing program
to determine the adverse effects before the
technology is widely available in general aviation. The
goal of exploring the impact of technological
determinism on general aviation was to understand

the affects so the aviation industry, academia,

aviation organizations, and the general aviation sector

can develop theoretical perspectives to reduce the

negative attributes associated new technologies.

CHAPTER II

REVIEW OF LITERATURE

The purpose of this qualitative single descriptive case study was to explore the experiences of general aviation pilots on the integration of TAA and to determine how the automation of TAA can be used to increase aviation safety and improve pilot decision-making skills. A review of the literature related to technological influences on integrating TAA in general aviation includes scholarly works from peer-reviewed journals, research studies, and articles pertaining the follow the areas: (a) leveraging technology to improve general aviation safety, (b) determining the effects on human factors, (c) improving flight training to address new technologies, (d) capitalizing on technological innovations to improve pilots' cognitive abilities, and (e) managing the integration of technology in general aviation. The review of existing literature highlighted the need to explore the integration of TAA in general aviation. The literature pertaining to technological innovations highlighted the effects of automation on general aviation; consequently, governmental, academia, and industry efforts to reduce the hazards are disjointed, too far-reaching, and lack measureable objectives.

Introducing new technologies in general aviation requires amendments to training requirements and government oversight, which the FAA struggles to maintain oversight on implemented technologies (Downer, 2010; USDOT, 2012). The general aviation sector needs strategic objectives to reduce the adverse effects of technological deterministic thinking on human factors, flight operations, training, cockpit designs and equipment, and regulatory oversight (Cassens et al., 2011; Halleran & Wiggins, 2010; Manzey, 2012; Oliver, 2011). The complexity of TAA requires constant evaluation of the cognitive and human factors impact on pilots to prevent automation and information overload (Lindo et al., 2012; Martins, Martins, Soares & Augusto, 2013; Rice, 2009). An additional implication was lack of regulatory oversight perpetuated problems since training and implementation of regulations lagged behind integration of new technology, which placed general aviation pilots at greater risk (Lindo et al., 2012; USDOT, 2012).

In 2008, there were 1,254 fixed-wing general aviation accidents, which 72% of accidents were due to pilot error (Nilsson, 2011). Integrating TAA in general aviation has resulted in unintended outcomes like accidents, fatalities, a critical training deficiency, and a lack of government oversight (Cassens, Young, & Greenan, 2011; McCracken, 2011, NTSB, 2012). Without comprehensive training programs, pilots are susceptible to misusing automation rather than leveraging the full scope of automated capabilities to capitalize on the advanced automation of TAA (Halleran & Wiggins, 2010; Mitchell et al, 2009). The missing dimension aimed at reducing general aviation accidents is the lack of scientific research that documents the adversity of technological innovations caused by technological determinist thinking. The general aviation sector needs a systematic process of integrating technologies without introducing new hazards and safety concerns (Di Renzo, 2010).

As of 2007, 90% of new general aviation aircraft have glass cockpits, which increased the number of

TAA in general aviation, to over 5,700 (NTSB, 2010).

Therefore, to target contemporary problems with TAA

in general aviation, a collaborative effort between the

FAA, academia, industry, private organization, and

general aviation pilots developed FITS (GAO, 2012;

Lindo et al., 2012). The primary objectives of FITS are

to improve general aviation pilots' proficiency at

operating TAA, maximizing utilization of the

automated capabilities, and enhancing pilots' higher

order of thinking through single pilot resource

management, aeronautical decision-making, and

situational awareness (Cassens et al., 2011).

The strategic goal is to curtail general aviation

accidents by 10% over a 10-year period (2009-2018);

thus, making this the first time the FAA established a

benchmark for general aviation (GAO, 2012).

Unfortunately, the FAA failed to achieve to a 1%

reduction in general aviation accidents in 2009, 2010,

2011, and 2012 (GAO, 2012). In 2010, the FAA

worked with the General Aviation Joint Steering

Committee (GAJSC) to implement a 5-year plan to

reduce general aviation accidents by enriching data to

develop accident mitigation strategies, increase

aviation outreach, promote safety, and implement risk

management procedures (GAO, 2012). The GAJSC is
a partnership between the federal government,
academia, aviation organizations, and the aviation
industry is charged with assessing general aviation
accident data to develop and implement preventative
strategies to reduce general aviation accidents (GAO,
2012). The data gathered from general aviation pilots
may lead to increased awareness on impacts of
technological innovations in general aviation.

The nature of technological deterministic beliefs
is to disregard the implications on society,
organizational culture, and the coercion of innovations
on consumers (Calvert, 2013; Jackson & Philip, 2010;
Lane & Lyle, 2011). The integration of technological
advancements have adversely impacted flight
training, reduced pilots' situational awareness,
reduced cockpit standardization, increased human
factors issues, and failed to reduced general aviation
accidents (Cassens et al., 2011; NTSB, 2010).
Positive aspects of TAA address safety and includes
(a) the advanced automated capabilities allow general
aviation pilots to fly farther, (b) familiarization with
glass cockpits during the early stages of flight training,
(c) comparable cockpit instrumentation to large jets,

(d) receive in-flight weather, access to automated data and autopilot capabilities, and (e) exposure to advanced flight training to safely operate TAA (GAO, 2012; Lindo et al., 2012; NTSB, 2010). Within general aviation, the effective integration of technologies has to account for automated functions, human factors, and human interface, for the most part; the focus is on leveraging the automated capabilities rather than on the negative effects (Pritchett, 2009). The implementation of new technologies without knowing the adverse impacts is not a prudent business practice, especially in this case, due to the FAA's strategic goal of targeting the reduction of general aviation accidents (GAO, 2012). Managers in the general aviation sector need to advocate for improved oversight and forecasting to safeguard existing safety practices. To understand the characteristics of technological deterministic thinking is essential for managers, especially since research indicated that the implementation of innovations is a common practice to resolve problems caused by technology (Calvert, 2013). The purpose of this study was to explore the integration of TAA in general aviation.

The Impacts of Technological Determinism.

Technological determinists deem technological development as a dynamism that influences the continuous integration of technological innovations that influences and shapes society (Chandler, 2012; Novicevic, Hayek, Buckley, & Humphreys, 2009; Yang 2009). Technological determinists believe the advancement of society would suffer without the continuous integration of new technological developments (Landeweerd et al., 2009). Chandler (2012) argued technologies are advantageous to humans by decreasing the physical and cognitive demands and increasing the availability of automated information even though human autonomy is restricted. Technological determinists assume innovations follow a course fundamental to the nature of the technology; as a result, more advanced technologies are replacing older innovations, and to interfere with technological innovations is to disturb the natural order of society (Novicevic et al., 2009). A principal belief of technological determinism is to disregard the social impacts of technological innovations and focus primarily on the technological advantages and the natural progression (Misa, 2009). Concentrating on the technological advantages and disregarding the adverse implications perpetuates the misconception that technology improves society.

Technological-based solutions are innovations designed to improve aviation safety or remedy previously identified problems such as innovations in ground and proximity warning systems, in-flight weather, terrain awareness capabilities, and TAA (Halleran & Wiggins, 2010; Oster, Strong, & Zorn, 2013). From 1980-2010, the aviation industry integrated technological-based solutions with an intent of improving safety; however, in some cases safety was actually threatened such as increased pilot workload, diminished situational awareness, atrophy of basic flying skills, and incompatibilities between aircraft and air traffic control systems (Archer, 2012; Doorn, 2012; McCloy, Harper-Sciarini, Durso, Jentsch, Kanki, & Rogers, 2011; Oster, Strong, & Zorn, 2013). These unintended consequences hindered aviation safety and efforts to reduce general aviation accidents and fatalities (Archer, 2012; NTSB,

2012). Members of the NTSB investigate over 1,600 general aviation accidents per year (Garibay & Young, 2013). Adverse impacts on cognitive and physical abilities of the abovementioned technologies are concomitant to technological deterministic thinking due to induced changes (Chandler, 2012; Oster, Strong, & Zorn, 2013). Technological determinists review induced changes as necessary deviations for improvements; however, the transition occurs without holistically understanding the negative impacts (Chandler, 2012). Methods to offset the negative attributes of innovations are through training (McCraken, 2011) and predictive analysis to forecast potential problems (Halladay, 2013).

Hughes, Rice, Trafimow, and Clayton (2009) declared that scientific research regarding automation focuses on the trust and reliance of the operator while disregarding consumers' viewpoints. This one-dimensional approach is indicative of how the aviation industry implemented technologies without knowing the systemic impact on the domain (King, 2011). In fact, the NTSB (2010) listed flight training, new equipment and operation, pilot performance, safety issues, and accident investigation methods as problems introduced by TAA. Technologically advanced aircraft have automated capabilities analogous to that of large jets, which researchers determined the automated features are difficult to comprehend without the proper training (Lindo et al., 2012; NTSB, 2010). The continuous integration of technological advancements in the aviation industry perpetuated increased pilot workload, outdated flight training paradigms, diminished situational awareness, atrophy of basic flying skills (Archer, 2012; Downer, 2012; NTSB, 2012). The continual integration of technologies challenges the FAA to provide regulatory oversight due to understaffing (Archer, 2012; Downer, 2012; NTSB, 2012).

Technological determinists support constant development of technologies for improvements; however, a negative impact of technological determinist thinking is the current regulatory system cannot keep pace with the rapid implementation of innovations (Chandler, 2012; Downer, 2012). Some technology determinists postulated technological influences are unidirectional, in which the negative outcomes are disregarded by technologists (Glenna, Jussaume, & Dawson, 2010). To reduce unintended consequences, researchers asserted that technology developers should consult with end-users to mitigate shortfalls and negative impacts (Lucivero, Swiersta, & Boenink, 2011). Therefore, technology developers need to consult with general aviation pilots to identify the negative impacts of TAA.

According to researchers, the level of automation in aviation affects human-automation interaction (HAI) (Hancock et al, 2013). Human-automation interaction involves decisions and inputs that pilots make when manipulating the automated capabilities of the aircraft. Lack of scientific research on HAI prevents researchers and practitioners from understanding the holistic effects on pilots (Hancock et al, 2013). Parasuraman and Manzey (2010) postulated automation is not a replacement for human activity; consequently, automated capabilities modify the interaction between human and machine that sometimes lead to degradation of technical skills due to complacency.

The largest stakeholder group in the aviation industry is general aviation pilots. The aviation industry integrated TAA in the general aviation sector for three reasons centric to technological deterministic beliefs: first, competition amongst industry rivals drove the implementation of advanced automation to enhance safety; second, the aviation industry relies on the advanced capabilities produced in commercial aviation; and third, to capitalize on the advanced capabilities of TAA (Chandler, 2012; GAO, 2012; NTSB, 2010). Ozdemir and Dermici (2012) postulated that leadership is a core capability of the innovative process because leaders are responsible for developing the strategic direction, communicating the objectives, and assuming operational ownership of the innovative process. One challenge for general aviation is the lack of a singular authority because the preponderance of general aviation activity consists of personal flying (GAO, 2012). General aviation is designed purposely in a flat hierarchical construct to allow pilots to conduct personal and recreational flight operations as long as all activities comply with federal aviation regulations. The lack of a hierarchical construct places the onus on general aviation pilots to understand the technological impacts caused by innovations. To offset the lack of a hierarchical construct, researchers recommended the creation of an incentive program to encourage pilots to undergo routine recurrent and safety training to increase safety for general aviation pilots (Garibay & Young, 2013).

Another area of concerns is implementing safety management systems in general aviation (NTSB, 2010) because the FAA is drafting a regulation mandating for each airline transport carrier to establish a Safety Management System (SMS) (Steckel, Lercel, Rieser, Kostal, & Patankar, 2013). The FAA is not mandating SMS for organizations or pilots in general aviation pilots (Steckel et al., 2013). Technological determinists emphasized that technological innovations influence change (Landeweer, Osseweijer, Kinderler, 2009). Even though technologies are intended to improve capabilities or processes; yet, unintended consequences like increased pilot workload, diminished situation awareness, atrophy of basic flying skills, and incompatibilities between aircraft and air traffic control systems emerge (Archer, 2012; Doorn, 2012; McCloy, Harper-Sciarini, Durso, Jentsch, Kanki, & Rogers, 2011). A SMS is designed to help identify safety vulnerabilities, weaknesses, and discrepancies (Steckel et al., 2013). Using a SMS in general aviation can help identify and eliminate

problems with technological advances (Garibay &
Young, 2013). Safety management system pertains to
developing a holistic culture and processes to identify,
prevent, and document methodologies to prevent
safety accidents (Lu, Young, Schreckengast, & Chen,
2011). Researchers believe a trickledown effect will
challenge the general aviation sector, especially for
pilots that own and operate aircraft for personal due to
the lack of motivation to design, implement, and
enforce SMS in general aviation (Steckel et al., 2013).
Leveraging SMS in general aviation will help
effectively manage the increasing use of innovations
in the general aviation sector by using a systematic
approach to identify hazards and emplacing corrective
measures to overcome the risks such as training
deficiencies, technology integration, and automation
management (Flouris & Yilmaz, 2009; Halleran &
Wiggins, 2010).

Managers and leaders have a moral obligation
to implement and utilize technological innovations that
ameliorate processes or procedures (Chandler,
2012). Because technological innovations improve
human capacity, managers and leaders have to
determine the appropriate uses of the innovations

(Chandler, 2012) to prevent unintended consequences (Doorn, 2010). Automated capabilities are prevalent in most domains; thus, increasing the scientific research and discourse on human automation interface enables practitioners to understand the appropriate use and levels of automation (Parasuraman & Manzey, 2010; Rice, 2009), and further revolutionize technology (Wright, 2011). Technological determinists focus primarily on the new automated capabilities of technological advancements; however, the lack of scientific data on methods to reduce complacency, human automation interaction problems, and increased workload remains a challenge for technologists and researchers (Parasuraman & Manzey, 2010; Rice, 2009). The effective implementation of technologies requires a well-balanced approach to capitalize on emerging capabilities while minimizing unintended consequences, (Chandler, 2012; Parasuraman & Manzey, 2010; Rice, 2009). The rapid integration of new technologies and staffing issues at the FAA prevent a more assertive effort to reduce unintended consequences (Downer, 2012). For example, the increasing complexity of civil aviation systems coupled with the mounting safety measurements are placing more demands and outpacing regulatory abilities at the regional, national, and international levels (Downer, 2012; Layton, 2012).

Some technological determinists focus on the integration of specific technologies while disregarding the peripheral impacts of new technologies (Batteau, 2010). The peripheral impacts are noticeable in the general aviation sector when examining the adversities of integrating TAA. These peripheral impacts in general aviation highlight regulatory, training, human-machine interaction, oversight, and misinterpretation issues, which is the result of poor technological integration and planning (Cassens et al., 2011; Hallström & Gyberg, 2011). To safeguard against unintended consequences of new technologies is a shared responsibility between researchers, technologists, and industry leaders to develop practices to mitigate negative consequences (Downer, 2012). The adverse impacts of technological determinism on general aviation are worth exploring because some practitioners and researchers believe the development of new technological innovations is the best approach to resolve existing problems

(Calvert, 2013). Based on Calvert's (2013) supposition, the continuous development of new technologies will continue to trouble general aviation because some practitioners and researchers believe the only way to overcome the problems with innovations is by implementing new technologies (GAO, 2012). Ninety percent of general aviation operations were personal and recreational flights (GAO, 2012), which places the cost of implementing and maintaining a SMS on individual pilots (GAO, 2012). The lack of an existing SMS construct in general aviation (GAO, 2012) prevents pilots from detecting issues with emerging technologies at the earliest opportunity (GAO, 2012). This is because general aviation pilots do not undergo safety management training to identify emerging problems.

Jackson and Philip (2010) stated "technology is the key determinant for driving cultural change and bringing about techno-change success and that change can be planned and implemented in a top-down fashion by senior management" (p. 446). A major disdain of technological deterministic thinking is the disregard for cultural sensitivity (Jackson & Philip, 2010). Not understanding the safety culture and safety initiatives in the general aviation sector is a significant quandary (GAO, 2012) because the implementation of new technologies creates training, regulatory, and human factors problems (NTSB, 2010). Human factor problems, outdated flight training, and the lack of regulations and oversight have been isolated to the integration of TAA (Archer, 2012; Lindo et al., 2012; Parasuraman & Manzey, 2010). The intent is to implement technologies to improve aviation safety; nonetheless, the innovations have resulted in more fatalities and aviation accidents (Garibay & Young, 2013) because the chief focus of technological advances is the influence and not the adverse effects (Chandler, 2012). The military and commercial aviation sectors have SMS and other layers of safety initiatives, which the general aviation sector lacks (Flouris & Yilmaz, 2009).

A problem with integrating innovations is the lack of theoretical understanding on the effects of automation on trust and human dependence due to the lack of knowledge on cognitive influences of automation and human reliance (Rice, 2009). Doorn (2012) argued for the inclusion of safety parameters, repeatable and sustainable practices, and the pilots' input during the technology development phase. Adhering to a strict implementation process will prevent a cascade of problems (Doorn, 2012) and allow the aviation industry and the FAA to counter the adverse effects of technological determinism.

Another consequence of technological deterministic thinking is automation surprise. The automated capabilities of TAA cause automation surprise (King, 2011). King postulated that automation surprise pertains to the increased suspicion, curiosity, and inquisitive concerns of how the systems operate, which leads to an antagonistic approach with the aircraft. Automation surprise is hazardous because this phenomenon affects inexperience and experience pilots; however, the training to reduce automation surprise in general aviation is unregulated, loosely organized, and not mandated (King, 2011). The integration of automation, flight management systems, and navigational capabilities are effective safety enhancers in the commercial aviation sector; however, there are no regulations to provide oversight on pilots transitioning to TAA (King, 2011; Lindo et al., 2012). The lack of mandated training have compelled aviation insurance companies to direct additional

training requirements to overcome the regulation and
training gaps and hazards like automation surprise
(King, 2011; Lindo et al., 2012). The technological
complexities of TAA such as dilapidated training
requirements, regulatory oversights, and human
factors issues require additional regulations to reduce
the number of general aviation accidents (Lindo,
2012; Whitehurst & Rantz, 2012). Members of the
NTSB (2010) recommended the FAA improve training
curriculums to equip general aviation pilots to make
informed decisions and to incorporate the automated
capabilities of TAA into the pilot's decision matrix.

King (2011) postulated the ascendancy of
automated systems in the air transport industry is
common; consequently, increasing the training
requirements for commercial pilots, yet automation
surprise occurrences continue to increase. The
increased occurrences of automation surprise in the
air transport industry are indicative of the let-downs in
the automated system designs and defective decision
matrices, which disregard the pilot's inputs (King,
2011). To overcome the faulty designs, King
recommended software upgrades that take the pilots'
inputs into account, which will improve decision-

making process. King recommended the general
aviation sector leverage corrective efforts from the air
transport industry to reduce automation surprise in
general aviation.

The integration of TAA in general aviation
reduced the pilot's role because glass cockpits
provide computerized data, advanced avionics, and
autopilot capabilities (Feary, McCloy, Wickens, Kaber,
Pritchett, & Sherry, 2010). General aviation pilots
flying TAA need to undergo FITS training to develop
an in-depth understanding of automated systems and
the automated information, specifically, how datum is
derived from the automation (Cassens et al., 2011).
The lack of pivotal regulatory guidance regarding
FITS is less than optimal because FITS is a
recommended training rather than a mandatory flight-
training requirement (Cassens et al., 2011). General
aviation pilots that have not completed FITS lack the
higher order thinking skills to operate TAA safely
(Cassens et al., 2011; Dimitriadis, 2012). Higher order
thinking skills are cognitive abilities that enable pilots
to make informed aeronautical and risk management
decisions, enhance situational awareness, and

increase task management and prioritization (Nilsson, 2011).

The implementation of new technologies is a strategy to improve aviation safety by reducing the number of general aviation accidents (GAO, 2012; Lindo et al, 2012). Aviation technologists lack in-depth understanding of how automation and new technologies affects safety (FAA, 2013). Aviation subject matter experts recommended new research on initiatives to reduce the negative aspects of aircraft automation so general aviation pilots can capitalize on the automated capabilities such as autopilot, global positioning systems (GPS) with moving maps, and advanced avionics (FAA, 2013; Franza & Fanjoy, 2012). Vanderburg (2012) stressed that aviation technologists and regulatory agencies need to understand the impacts of implementing technologies because innovations create new problems, which affect all sectors of aviation, including general aviation. These problems result in complacency and human factors that cause accidents (NTSB, 2010). Aviation experts avowed the implementation of new technologies require corresponding regulatory

changes to safeguard against the hazardous impacts
(FAA, 2013).

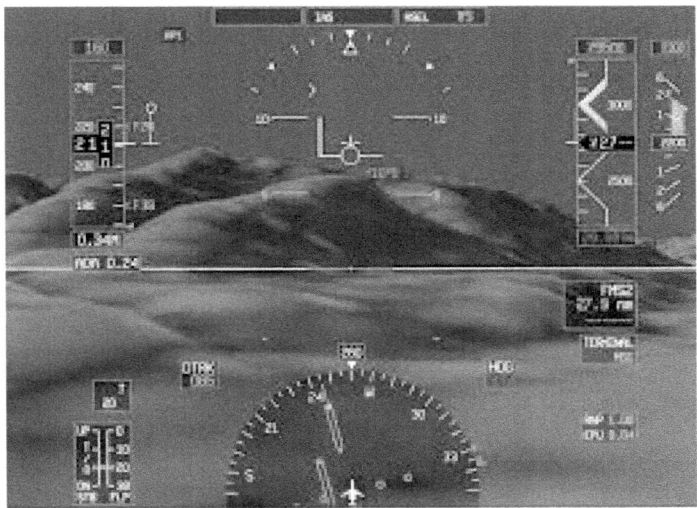

*Figure 9 - An image of Honeywell's synthetic view for pilots;
photo courtesy of the Honeywell Corporation.*

The negativism of technological advancements
center on the social and organizational cultural
implications that resulted from unintended
consequences that caused accidents (Chandler,
2012; Parasuraman & Manzey, 2010). The aviation
industry continuously implements new capabilities
regardless of the safety, political, and institutional
implications (GAO, 2012; Misa, 2009; Yang, 2009).
While the quality of automation in TAA continues to

improve; nonetheless, unanticipated problems and failures with the integration of new technologies continue to affect human automation interface (Friday, 2011). Regardless of the changes implemented to improve automation, technology developers cannot account for human behavior because automation changes the humanistic responses and tasks (Parasuraman & Manzey, 2010).

Friday (2011) stressed that inexperienced general aviation pilots are vulnerable to fixation; therefore, pilots are prone to make inappropriate decisions when manipulating the autopilot, which is an example of automation complacency (Parasuraman & Manzey, 2010). Pilots that fly TAA are prone to become over reliant on automation, which diminishes basic piloting skills (Friday, 2011). From a political and policy perspective, the FAA has not made any significant changes to regulations since 1977; yet, the proliferation of TAA in general aviation is prevalent (Friday, 2011). Technologically advanced aircraft require a different approach to flight training, which is non-standardized and problematic because of the various avionic systems available (Friday, 2011).

The implementation of technology in general aviation is not without safety, societal and institutional challenges, which is evident by the FAA developing and recommending FAA and Industry Training Standards (FITS) for general aviation pilots that are transitioning to TAA (Halleran & Wiggins, 2010). The purpose of FITS is to provide general aviation pilots a comprehensive flight-training program to offset potential increases in accidents caused by the complexities of TAA (Halleran & Wiggins, 2010). The FITS Program is a cultural, institutional, and societal change to traditional flight-training programs (Chandler, 2012; Halleran & Wiggins, 2010). The FITS Program brought about institutional changes such as training general aviation pilots to fly complex aircraft like commercial pilots (Lindo et al., 2012). The FITS Program is a course to teach general aviation pilots new skills and to think differently by expounding on different scenarios when operating TAA aircraft. A widespread misconception about new technologies in general aviation is that innovations will reduce the number of aviation accidents (Halleran & Wiggins, 2010). Integrating TAA in general aviation has not reduced the number of accidents; consequently, researchers have determined that TAA introduced new problems in general aviation (GAO, 2012; NTSB, 2010).

Some practitioners believe automation is desirable, because it reduces the pilot's workload and improves cognitive ability by providing pilots with automated data unlike conventional aircraft (Pritchett, 2009). Some researchers are apprehensive about TAA because of the negative effects associated with the glass cockpits (Pritchett, 2009). Some problems with TAA include the lack of mandated training, human automation interaction, the ineffectiveness of FITS, and the inability to reduce automation surprise in general aviation (King, 2011). Gray (2009) stated the aviation industry leveraged new technologies to improve effectiveness, which in turn increased profitability. The technological advantages of automation are the gravitational pull that drives the continuous integration of technology to reduce general aviation accidents, which unsystematically leads to implementing more technological capable aircraft that further exacerbates the problems (Cassens et al., 2011; Lindo et al., 2012; Misa, 2009; Pritchett, 2009).

The emergence of new technology into society appears to be transparent; however, according to Yang (2009) there are always adverse effects associated with new technologies. Technological deterministic thinking adversely affects the activities and behaviors of humans and organizations because innovations cause a societal shift from normal patterns, in which some researchers argue that technological change is not mutual especially to those organizations or people opposing change (Douglas, 2010; Misa, 2009). Researchers concluded that technological innovations affect policies, the populace, and cultures because technology driven organizations are focusing on the capitalistic gain rather than the impact on politics, normative, and cognitive influences of innovations. The divergence between technological organizations, the populace, and researchers occurs because historians, researchers, and anthropologists have different interests; therefore, it creates an educational chasm and misconception (Misa, 2009). Understanding the long-term effects of technological determinism assists with refocusing the underpinnings of the theory.

Critics of technological determinism stress the adverse implications some innovations have on strategic decision-making, societal, and institutional factors, which are surpassed by the capitalistic gains of new technologies (Ross, 2009). Some researchers emphasize the progression of technological developments is a natural order while antagonists argue humans have authority and power to regulate implementation of technology (Novicevic et al., 2009).

*Figure 10 - Current and prototype components of Coast Guard
HC-130 Hercules legacy cockpit complete with upgraded flight
displays, upgraded weather radar and yoke-mounted Garmin
GPS MAP 695/696 portable aviation receivers at the Aviation
Logistics Center in Elizabeth City, N.C., Oct. 21, 2010; photo
courtesy of the U.S. Coast Guard and Lt. Cmdr. Brian Erickson.*

Doz (2011) and Liang (2010), acknowledged lack of theory and academic research in the business domain, which is further complicated by technological deterministic beliefs that affect multiple domains; thus, making it difficult to explore the impact of technological deterministic concepts. For example, general aviation is an area lacking scientific research; therefore, research conducted by Halleran and Wiggins (2010) and Cassens et al., (2011) explored specific aviation studies that highlight the adverse impacts of TAA, which is indicative that industry stakeholders are willing to accept the negative attributes, because the significance of the innovation outweighs the adverse associated problems. Ehrmann (2011) suggested the implementation of technology as a mechanism for change without transitioning too fast from one technology to next because innovations can enrich the process. Transitioning too fast disrupts collaborative efforts and interrupts the standardized methodologies (Ehrmann, 2011), which is what the general aviation sector is experiencing.

The current methodologies of conducting research on technology will not provide substantial details to indicate a need for a change because there is a requirement to reduce the literature chasm "between empirical studies of technology and philosophical analyses of modernity" (Misa, 2009, p. 358). The existing literature gap allows essential questions to remain unanswered because scientists and practitioners have not demonstrated the propensity to explore the sociological, psychological, political, or moral domains to determine the actual impact of technology on the society (Misa, 2009).

General aviation maintenance is a relevant area of concern due to technological development and integration. According to Pettersen, McDonald, and Engen (2010), in the aviation industry, human error is the primary reason for accidents in empirical studies; consequently, aviation maintenance is a vital linkage in the system of systems approach in aviation. In fact, 7% of general aviation accidents were a result of maintenance related problems (Reynolds, Blickensderfer, Martin, Rossignon, & Maleski, 2010). Maintenance processes and procedures changed due to the integration of advanced digital equipment in the aviation industry (Gray, 2009). The integration of technologies in the aviation industry created a shortage of proficient aviation maintenance personnel that specialized in repairing advanced technologies (Gray, 2009). Aviation maintenance personnel require the technical acumen and capacity to maintain computers, optics, artificial intelligence, and sensors, which are integral components of the avionic suite in TAA (Gray, 2009). Leaders in the aviation industry need to scrutinize the feasibility of all technologies (Gray, 2009) because some technologies have unintended consequences (Downer, 2012) and negatively affects productivity and efficiency (Gray, 2009).

The cross-domain impacts of technological deterministic beliefs affect multiple domains because aviation is a socio-technical system that makes aviation susceptible to organizational learning and leadership implications (Craig, 2012; Gray, 2009; Parchoma, 2009; Ropp, 2009). Implementing technology is deemed necessary to maintain the competitive advantage for safety concerns; however, Hendrickson (2009) articulated some technologies created new problems, especially with the government struggling to regulate technological compliance, which led to formidable training problems

that were disregarded (Downer, 2012; King, 2011).
These oversights led to fatalities and aviation
accidents in the general aviation sector (GAO, 2012)
because business practitioners, technology
developers, and organizational leaders do not
understand the systemic implications of innovations.
A conventional paradigm in safety management is to
focus on a singular incident and implement corrective
actions (Smith, 2011) rather than holistically assess
the impact on the system of systems construct (Craig,
2012). Smith (2011) recommended safety analysts
move beyond using analysis to understand impacts
on systems and leverage synthesis, a processing of
evaluating each component or layer individually.
Some system engineers believe using analysis-based
research will correct unsafe actions; however, to
resolve unintended consequences a complete review
and breakdown of each component in the system is
necessary to prevent unsafe acts (Smith, 2011).

General Aviation Initiatives.

The FAA is a leading advocate of FITS to
prepare general aviation pilots to operate advanced

and technically capable aircraft by teaching pilots
aeronautical decision-making, single pilot resource
management, and task prioritization (Cassens et al.,
2011; NTSB, 2010). General aviation pilots can
capitalize on FITS and the lessons learned from
airline pilots who encounter analogous vulnerabilities
when transitioning aircraft to glass cockpits (NTSB,
2010). Black and Chimka (2011) asserted, "general
aviation is regulated differently than the commercial
aviation industry, making it difficult to borrow
improvements" (p. 47). The training methods used by
the airline industry for pilots upgrading to advanced
aircraft are worth consideration to develop a similar
program for general aviation pilots. Researchers,
practitioners, and managers are aware of the difficulty
in leveraging capabilities from the commercial or
military aviation sectors; yet, when the same
innovations are integrated in the general aviation
sector – new problems surface (Pritchett, 2009). To
capitalize on the technological capabilities of TAA
requires general aviation pilots to shift from outdated
training paradigms to training programs like FITS
(Halleran & Wiggins, 2010; Lindo et al., 2012). Using
FITS exposes general aviation pilots to operational

scenarios and malfunctions that could occur in glass cockpits, which builds confidence and enhances the aeronautical decision-making skills of general aviation pilots (Halleran & Wiggins, 2010).

Aeronautical decision-making is a process to improve the pilots' cognitive and problem solving skills; therefore, enhancing the pilots' judgment, higher order thinking, and mental acumen when affronted with a particular situation (Cassens et al., 2011). The integration of TAA in general aviation forces pilots to transition from traditional training paradigms to account for more advanced aircraft, as an effort to reduce general aviation accidents (Cassens et al., 2011). Because general aviation pilots are flying TAA, which are technologically comparable to commercial and military aircraft, then the general aviation sector needs to train its pilots using similar curriculums to reduce the probability of general aviation pilots having accidents in TAA.

To reduce the number of general aviation accidents caused by pilot error, the FAA along with the aviation industry, recommended FITS (Cassens et al., 2011). One purpose of FITS is to capitalize on scenario-based training (SBT), a methodology that

incorporates real world scenarios to teach
aeronautical decision-making, risk management, and
single-pilot resource management skills during flight
operations (Cassens et al., 2011; Halleran & Wiggins,
2010). One objective of FITS is to enhance the
aeronautical decision-making of general aviation
pilots to lower the number of accidents (Hendrickson,
2009).

The development of new technologies has
lingering implications beyond social and cultural
impacts (Chandler, 2012). Researchers indicated
some technologies like digital technology defy existing
laws and legal policies (Cockfield, 2011). This is
indicative of technological innovations outpacing the
legal the system (Cockfield, 2011). Antagonists of
technological determinists argue for humans to
interpose on technologies to prevent undesirable
consequences of technologies (Chandler, 2012). The
impact on the legal system is an unintended
consequence that researchers, practitioners, and
industry leaders need to resolve before implementing
technologies. For instance, existing literature does not
address the impact of the Next Generation of Air
Transportation System (NextGen) on general aviation.

The NextGen an emerging state-of-the-art capability to improve air transportation by ameliorating safety and decreasing environmental implications are an example of the FAA using technology to improve existing conditions in managing air traffic (Sawyer, Berry, & Blanding, 2011). Yet, researchers recognize the need to identify the positive and negative aspects of NextGen to reduce the impacts on human performance and national airspace system (Sawyer et al., 2011). Assessing the overall impacts of NextGen in advance illustrates the FAA's efforts to minimize the technological determinism impacts by decreasing the social and cultural affects (Chandler, 2012) and implementing an innovative capability necessary to overcome problems extending from a dilapidating airspace architecture (Sawyer et al., 2011). The implementation of NextGen is an effort to improve air traffic control operations and safety (Sawyer et al., 2011) and the general aviation sector can benefit from NextGen. The FAA, academia, general aviation organizations, and general aviation pilots need to develop a roadmap to address the impacts of NextGen and the required training for pilots flying within the NextGen infrastructure.

Another effort to improve general aviation is through the Safer Skies agenda, which is an initiative that targets reducing the number of general aviation accidents (GAO, 2012). According to the GAO (2012), the Safer Skies Strategy includes two groups, the General Aviation Joint Steering Committee (GAJSC) and FITS, which are charged with combating general aviation accidents by using SBT. The TAA Team, which is a subordinate of the GAJSC, is primary responsible for focusing on the unique training and safety problems stemming from the integration of glass cockpits in general aviation (GAO, 2012). This team worked in conjunction with general aviation pilots, academia, the FAA, and other aviation organizations to tackle reducing general aviation accidents. In a unified effort, members of GAJSC and FITS impressed on the general aviation community, that failure to undergo FITS training before operating a TAA is counterproductive to reducing general aviation accidents (GAO, 2012).

The benefit of implementing FITS in the general aviation sector is to change the pilots' perspectives of operating TAA and to instruct low-time or student pilots to adapt to an automated environment and use higher order thinking skills analogous to methods taught to airline pilots (Halleran & Wiggins, 2010). The TAA Automation Subgroup of the GAJSC is responsible with determining the specifics of FITS based on the levels of automated capabilities of the aircraft (GAO, 2012). The key to determining the effectiveness of FITS and the GAJSC is dependent on future trends in general aviation.

The general aviation sector is teaching single pilot resource management to pilots that is analogous to crew resource management, a strategy used by military and commercial platforms with multiple crewmembers (Cassens et al., 2011). In single pilot resource management, the pilot is the sole crewperson responsible for managing the internal and external factors of the flight (Cassens et al., 2011). The purpose of single pilot resource management is to teach the pilot to work methodically through a situation using aeronautical decision-making and minimizing risk by the making more informed decisions (Cassens et al., 2011). Single pilot resource management is significant for general aviation pilots as sole manipulators of TAA because it teaches the pilot to use to a systematic and deliberate process in the cockpit. The pilot needs to understand the intricate workings of the advanced technologies to capitalize on single pilot resource management and the aircraft's automated capabilities.

Leveraging Technology to Improve Safety.

The integration of TAA in general aviation is problematic and indicative of the unique challenges of leveraging new technologies to improve safety in general aviation (Lindo et al., 2012). The problems associated with TAA in general aviation are: (a) outdated flight training practices, (b) the nonexistence of a joint government and non-government organization to regulate technologies in general aviation, (c) no defined TAA proficiency training requirements, and (d) the lack of a requirement to teach resource management skills to general aviation pilots (Franza & Fanjoy, 2012). An unintended consequence of integrating TAA in general aviation is the requirement to teach single pilot resource management to general aviation pilots (Halleran & Wiggins, 2010). Chandler (2012) postulated an education implementation such as single pilot resource management is a social impact because of the widespread change to the existing curriculum, which stems from technological deterministic thinking. With this TAA integration, new requirements mandate that certified flight instructors teach single piloted resource management to general aviation pilots (Lindo et al., 2012). In a NTSB (2011) report, the data indicated that increased automation changed the pilot's workload from task performance, which means less flying the aircraft and more monitoring of systems, which places more emphasis on higher-level cognitive tasks. The commercial airline industry identified similar problems with increased automation in earlier studies; yet, the general aviation sector failed to implement measures to prevent unintended consequences with automation in general aviation (Halleran & Wiggin, 2010; Mitchell, Vermuelen, & Naidoo, 2009). The failure to capitalize on the existing problems in the commercial airline industry is indicative of the dogmatic nature of technological deterministic beliefs and the disregard for researching the unintended consequences (Chandler, 2012; Cockfield, 2011). Lindo et al. (2012) declared the lack of standardized practices and inability to leverage the advanced capabilities is problematic and requires immediate attention from academia and industry experts.

Figure 11 – JAL Boeing 747-400, Flight Deck; Photo courtesy of Norio Nakayama, Saitama, Japan.

The aviation industry seeks to improve safety by implementing technological innovations like glass cockpits, enhanced avionics systems, and automated air traffic control equipment (Di Renzo, 2010; Lee, Jeon, & Choi, 2012). The aviation industry attempted to capitalize on integrating TAA to improve general aviation safety by enhancing pilots' situational awareness; however, new problems arise with the implementation of innovations (Di Renzo, 2010; Lee,

Jeon, & Choi, 2012). Researchers, regulatory officials,

and the aviation industry underestimated the impact

of technological developments on general aviation;

consequently, the miscalculations led to challenges

with (a) training, (b) safety (c) management, (d)

standardization, (e) integration, and (f) human factors

(Franza and Fanjoy, 2012; Mitchell, et al. 2009;

NTSB, 2010). The goal of implementing TAA in

general aviation was to reduce the number of

accidents because 70% of general aviation accidents

were caused by pilot error (GAO, 2012; 2010; Sharma

et al., 2009).

The FAA is estimating there will be 750,000

general aviation pilots by 2016 and with the general

aviation sector having the highest number of

accidents, the onus is on general aviation pilots, the

FAA, academia, and aviation organizations to

implement measures to curtail the number of

accidents occurring (Bolstad, Endsley, Costello, &

Howell, 2010). The Air Safety Foundation linked the

primary causal factors of general aviation accidents to

poor aeronautical decision-making (ADM), lack of

situational awareness, and the lack of proficiency

(Bolstad et al, 2010). In a Government Accountability Office (GAO [2010]) report, the analysis indicated general aviation pilots are not required to undergo annual training on a regular basis; however, the pilots must complete biennial reviews with a certified flight instructor. With the integration of TAA and the need to modify flight training, researchers indicated current flight training paradigms are not effective (GAO, 2102). The integration of TAA accompanied by a growing pilot population may lead to an increase in general aviation accidents (GAO, 2012).

By establishing the Aviation Safety Program (AvSP), the FAA and National Aeronautics and Space Administration (NASA) identified information technologies to ameliorate the aviation architecture by codifying data to evaluate trends that are hazardous to aviation (Sharma et al., 2009). The AvSP strategy led to the development of an aviation technology database that consists of aeronautical innovations to improve the National Airspace System (NAS) (Sharma et al., 2009). The AvSP technologies are innovations that may have immediate impact by reducing aviation risks (Oztekin & Luxhøj, 2010).

The AvSP assisted in the coordination,
prevention, intervention on aviation issues, and the
reduction of adverse influences of new technologies
(Sharma et al., 2009). NASA maintains a portfolio of
new technologies to research the innovative
effectiveness of each capability Invented to reduce
aviation accidents (Sharma et al., 2009). Analysts at
NASA predicted for the aviation industry to improve its
safety standards, the integration of new technologies
is necessary (Sharma et al., 2009). The integration of
new technologies requires oversight and direction
from the FAA; however, the process is experiencing
delays to due to the lack of guidance on emerging
technologies in which the FAA is charged with
establishing policy, regulations, and authorizing
technological changes (Yeh, Swider, Abbott,
Donovan, Neiderman, & Piccione, 2012).

The influences of technological changes
facilitated aviation stakeholders in implementing
NASA's strategic plan: (a) to identify and integrate
innovations to improve aviation safety for aircraft to
operate in the NAS infrastructure, (b) to acquire, test,
and develop new aviation capabilities due to
substantial increases in air traffic, and (c) to

ameliorate existing architectures to optimize aviation system (Sharma et al., 2009). A common theme of extant literature places emphasis on the need to implement new aviation technologies to enhance general aviation safety. The challenge is integrating new capabilities without introducing new problems, FAA's human factors specialists work with the aviation industry to determine the design, to conduct the evaluation, and to counterbalance emerging innovations that require modifications to regulations, policies, or governmental direction (Yeh et al., 2012).

According to Sharma et al. (2009), "safety seeks to avert unintentional life-threatening events, while safety research in the AvSP involves developing prevention, intervention, and mitigation technologies and strategies aimed at one or more causal, contributory, or circumstantial factors associated with aviation accidents" (p. 843). As a commercial licensed pilot and former Aviation Safety Officer, one can stress the significance of reducing risks and hazards due to aviation being inherently dangerous. Due to aviation being a system of systems, it demands a complete comprehension of the interrelatedness and interconnectedness of the many sub-systems and the

influence of each system on the functionality to
minimize risk and hazardous (Craig, 2012; Ropp,
2009).

Figure 12 - Six basic instruments in a light twin-engine
airplane arranged in a "basic-T"; from top left:
airspeed indicator, attitude indicator, altimeter, turn
coordinator, heading indicator, and vertical speed
indicator; photo courtesy of Meggar.

With the integration of technologies, aviation
stakeholders rely on the AvSP to forecast the safety
effectiveness of innovations to minimize the adverse
effects (Mitchell et al., 2009). The economic and
operational improvements of new technologies
outweighed the adverse effects that the innovations

have on the pilots (Mitchell at al., 2009). New

technologies will continue to impact human factors

unless aviation leaders take comprehensive

measures to protect the NAS from the negative

effects (Mitchell et al., 2009). To determine the

hazardous implications of new technology on the

general aviation sector requires a holistic approach

and a strategic alliance between "pilots, air traffic

controllers, technical operations personnel, flight

attendants, and maintenance technicians all involved

company leaders, labor leaders, scientists, and

regulators in study design, data collection,

implementation of findings, and development of

products" (Aver & Johnson, 2011, p. 91). Avers and

Johnson postulated that stakeholders should leverage

existing scientific data and best practices developed

across the aviation domain to ameliorate safety

conditions to counter hazards. The missing variables

in the equation are the best business practices for

innovations, assessments of the implemented

technologies in general aviation, and a methodology

to select technologies. Rivalry competition,

exponential technological expansions, increasing

consumer demands, and amendments to government

regulations are drivers that increase development of new technologies (Das, Kumar, & Kumar, 2011).

In a study conducted by Halleran and Wiggins (2010), most pilots preferred TAA due to the automation that reduced the pilot's workload and the advanced avionics suite, which increased situational awareness. The research participants of the aforementioned study focused on airline pilots rather than general aviation pilots. Due to the integration of technologies within the aviation socio-technical construct, researchers are apprehensive about the impact of innovations on human factors, work overload, and situation awareness (Lee, Jeon, & Choi, 2012). To capitalize on the advanced capabilities of TAA, Mavin and Murray (2010) recommended the use of flight simulators because of the replicate cockpit designs and the simulated flight characteristics. Flight simulators are cost effective and enhance aviation safety by allowing pilots to practice flight maneuvers and procedures that are too risky to perform in the aircraft (Mavin & Murray, 2010). In the aviation industry, simulators are used as a practice-based method to improve skill level (Mavin & Murray, 2010). Some researchers argued flight training in simulators should occur in the exact replicas of the aircraft to achieve effective learning transfer; however, this supposition lacks support from empirical research (Mavin & Murray, 2010).

Researchers claimed advancements in aircraft technologies have created gaps in contemporary flight training programs, which is problematic and requires training objectives to address communication, aeronautical decision-making, and single pilot resource management (Mavin & Murray, 2010), especially in general aviation.

Determining the Effects on Human Factors.

Human factors are variables that influence human-machine interaction that may lead to less than optimum execution or degradation of human performance, which include systems, processes, procedurals, equipment, and social and psychological conditions (Craig, 2012; Paletz, Bearman, Orasanu & Holbrook, 2009). Human factors experts use a model known the Human Factors Analysis and Classification System to categorized aviation accidents to determine the human discrepancies that contribute to an accident (Paletz et al., 2009). Knowing that aviation maintenance contributed to general aviation accidents, government and industry authorities have not directed or requested that general aviation maintenance personnel undergo human factors training (Reynolds et al., 2010). The International Civil Aviation Organization, of which the United States is a member, mandated all aviation maintenance personnel in the commercial aviation sector undergo human factors training (Reynolds et al., 2010).

Air traffic operations are a vital part of the socio-technical system of aviation (Craig, 2012; Lee, Jeon, & Choi, 2012); therefore, the technological changes in air traffic operations affect procedures, equipment, and policy (Lee, Jeon, & Choi, 2012). Lee, Jeon, and Choi declared technological innovations in air control operations have advantages and disadvantages that affected human factors, trustworthiness, and safety, because pilot error is the primary causal factor in aviation accidents. Researchers are focusing on human error rather than technical miscalculations due to the systemic impact on the socio-technical system because technological innovations are becoming increasingly complex (Lee, Jeon, & Choi, 2012).

Technological innovations that are improving air traffic operations are introducing human factors problems for designers, decision-makers, and maintenance workers (Lee, Jeon, & Choi, 2012). National and international regulatory agencies implemented regulations to ameliorate navigational aids, airspace control, air traffic management, and emergency protocols coupled with the demand to handle more air traffic safely and efficiently (Lee, Jeon, & Choi, 2012). To accommodate the air traffic demands, the FAA authorized the implementation of technological innovations; thus, bringing about new problems that disrupted the aviation socio-technical system (Lee, Jeon, & Choi, 2012; Smith, 2011). Researchers noted that automation in air traffic control needs in-depth scrutiny to lessen the disruption of existing operations, safety practices, workload, and situation awareness (Lee, Jeon, & Choi, 2012). A major concern stemming from implementing air traffic technologies is the possibility of impeding air traffic controllers' situation awareness, which jeopardizes the safety of pilots (Lee, Jeon, & Choi, 2012).

Another area of concern is organizational ergonomics, which involves the maximization of the

socio-technical system by determining the criticality of
the components, processes, policies, and designs
(Cimpian, 2011). Organizational ergonomics is
reactive ergonomics, which is a reactionary approach
to correct a discrepancy that impedes the socio-
technical systems (Cimpian, 2011). The introduction
of technological innovations has unforeseen
consequences that forces the aviation industry into
reactive ergonomics because each innovation has the
potential to cripple the aviation socio-technical system
(Cimpian, 2011; Smith, 2011). To prevent
technologies from affecting the socio-technical system
requires a shift from analysis-based assessments to a
synthesized process to allow researchers to predict
the impact before implementing the technology
(Smith, 2011).

The rapid pace of technological integration in
the aviation industry threatens existing safety
practices due to increasing challenges of
understanding the impact of technology on aviation
(DiRenzo, 2009; Lee, Jeon, & Choi, 2012).
Government authorities have not mandated predictive
analysis, which would oblige human factors
specialists to evaluate the adverse impacts before

widespread integration (Berry, Stringfellow, &
Shappell, 2009). An unsettling find in existing
literature is integration of aviation technologies has
led to more accidents (GAO, 2012). Hendrickson
(2009) declared as automation increased, so did the
number of aviation accidents. General aviation
accidents that involved fatalities were four times
higher to be associated with a major contravention
when compared to non-fatal accidents (Berry et al.,
2009). The ascendancy of automation in aviation has
resulted in inexperienced pilot errors misusing the
automation, unfamiliarity with new technology, and
overreliance on the advanced capabilities
(Hendrickson, 2009; Pritchett, 2009). One impact of
technological deterministic thinking in general aviation
is the need to shift flight training from being
proficiency oriented to primarily focusing on
aeronautical decision-making, situational awareness,
and managing automation (Cassens et al., 2011).

Advanced automation is becoming a common
capability in general aviation aircraft; consequently,
testing and stressing the cognitive abilities of general
aviation pilots because pilots are required to manage
the multitasking and mentally understand the data

provided by the aircraft advanced instrumentation
(Taylor et al., 2013). Taylor et al. (2013)
recommended system engineers annotate the
cognitive demands of each task so system designers
can determine the level of automation required to
perform the task. Understanding the level of
automation needed for each task prevents misuse,
disuse, and the atrophy of critical skills (Taylor et al.,
2013). Allowing operators the ability to manipulate the
level of automation is known as adaptable automation
(Taylor et al., 2013). Adaptive automation enables
pilots to develop nontechnical skills such as
communication, aeronautical decision-making,
situation awareness, and task management (Kearns
& Sutton, 2013) comparable with the pilots' cognitive
comprehension of automation. Pritchett's (2009)
empirical work determined automation involves more
than off-loading tasks, because operators need to
understand intricate functions and dynamics of the
systems and the automated data provided, which
affects the human automation interface. Automation
scholars believe the complexities of automated
capabilities cannot be transferred by training,
procedural implementation, or human automation

interfacing; therefore, additional research is
necessary to develop a systematic and effective
approach for managing automation (Pritchett, 2009).

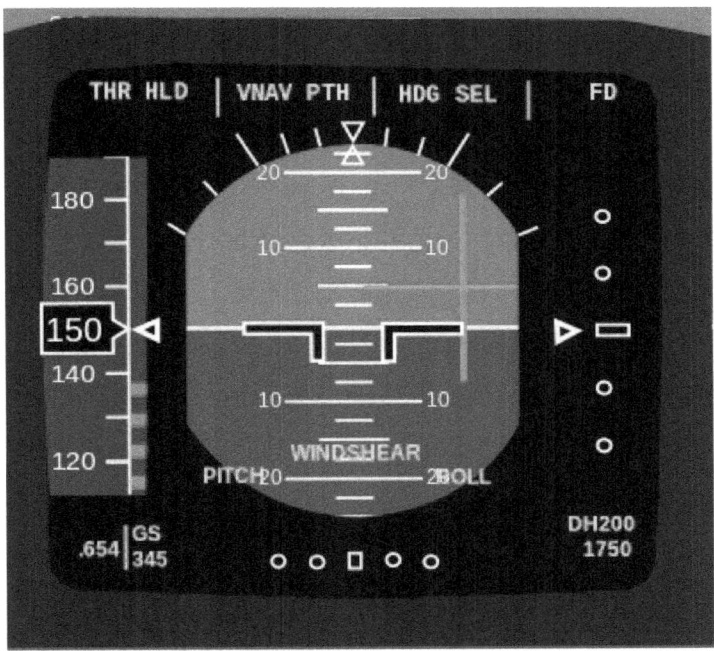

Figure 13 - Auto-pilot Panel; Photo courtesy of D-Laser.

The automated cockpits initially started with
capabilities to maintain heading and altitude (Pritchett,
2009). Since the beginning of flight, general aviation
pilot training was based on conventional or analog
instrumentation (Smith, 2008). Pilots learned to

maintained situation awareness, navigational and directional control, and practices to manipulate the aircraft according to the analog instrumentation (Smith, 2008). The evolution of TAA and glass cockpits initially started in the 1960s with the development and implementation of cathode ray tube (CRT) displays that replaced the analog gauges in the cockpit (NTSB, 2010). The use of CRT started the transition from analog gauges and equipment to glass cockpits because of the liquid crystal displays (NTSB, 2010). In 1974, NASA tested the utilization and feasibility of glass cockpits in commercial platforms, which led to the adoption of TAA to reduce the flight crew's workload through advanced automation (NTSB, 2010). From the 1980s, earlier renditions of glass cockpits started to appear in general aviation aircraft (NTSB, 2010). Another technological upgrade to the earlier renditions of glass cockpit was the implementation of flight management systems (FMS). The FMS enabled flight crews to achieve navigational precision because FMS automated the horizontal and vertical direction control of aircraft movements along a directed flight path (Mosier et al., 2013). The integration of FMS further automated the cockpit;

hence, forcing pilots to become information managers, which led to complacency (Casner and Schooler, 2014). Casner and Schooler posited that complacency was the result of task unrelated thoughts due to extensive automation in the cockpit.

Systemic human machine interaction and aircraft control issues of glass cockpits increased significantly between the 1980s and 1990s (NTSB, 2010). Pilots transitioning from conventional aircraft to TAA experienced increased workloads during emergencies, information management overload, cockpit scan issues, overreliance on automation, and degradation of basic piloting skills (McCracken, 2010; NTSB, 2010). A contemporary threat to general aviation is the various models of TAA, which is compounded by limited scientific research on the impact of glass cockpits in general aviation (NTSB, 2010; Robertson, 2010). Safety experts at the NTSB (2010) acknowledged that general aviation aircraft are outfitted with the same automated capabilities as military aircraft and commercial airliners. Technologically advanced aircraft were designed to reduce pilot workload; consequently, the integration of glass cockpits has redistributed the workload

(McCracken, 2011) to information management

(NTSB, 2010). According to McCracken, additional

studies are needed to determine the impact of virtual

data on general aviation pilots.

With the ascendancy of TAA in general aviation,

conventional analog instruments were replaced with

primary function and multifunction displays that

replace the airspeed, altimeter, heading indicator, turn

coordinator, and altitude indicator instruments (NTSB,

2010). McCracken (2011) postulated that transitioning

from conventional aircraft to TAA is challenging due to

the lack of familiarization with cockpit instruments,

colors, and symbols of the advanced avionics. For

pilots with considerable experience in TAA,

researchers are concerned with complacency, the

cognitive and decision-making processes of pilots,

and automation surprise (Geiselman, Johnson, &

Buck, 2013). One researcher acknowledged the

cockpit scanning problems for general aviation pilot

transitioning to the TAA because traditional scan

patterns used in conventional aircraft are not

applicable to TAA (McCracken, 2011). To overcome

the scan differences, general aviation pilots have to

learn the cockpit layout and interpret the data

provided on the primary and multifunctional displays (McCracken, 2011). McCracken emphasized the need for additional studies on general aviation pilots transitioning to TAA because more aircraft manufacturers are developing aircraft with glass cockpits. McCracken stressed the difference between conventional aircraft and TAA scan procedures, which is problematic because there are no formalized scan procedures for instructing such techniques in TAA.

A prominent concern with the integration of TAA in general aviation is to understand the human factors impact. Information management, automation management, complacency, instrumentation scan, inconsistency of glass cockpit designs, lack of cockpit familiarization, outdated training programs, and an increase in general aviation accident fatalities are compelling issues affecting human factors (McCracken, 2011; NTSB, 2010; Smith, 2008). These problems stem from integration of TAA in general aviation, which are compounded by lack of scientific research. The lack of scientific research on TAA in general aviation impedes the process of alleviating problems due to integration of TAA in general aviation.

Some researchers criticized technologies that targeted aviation weather because the innovations lacked human centered design features (Craig, 2012). Essential aspects of designing aviation technologies are integration and the level of intuitiveness from the pilot's perspective given that automated data and the multi-function and primary function displays are resources that provide graphic and informational details; therefore, avionic designers have to understand the effects on pilots (Craig, 2012). Because aviation weather contributes to 23% of accidents, practitioners in the aviation industry are installing near real-time weather displays in cockpits to enhance situation awareness (Craig, 2012). Human centered design is a concept to reduce the complicatedness of automated data systems and increase the intuitiveness for pilots; therefore, the FAA's Flight Deck Human Factor Research Program provides guidance in the development of technologies to avert complicated systems from affecting flight systems and applications (Craig, 2012; Yet et al., 2012). The FAA estimated that air traffic would increase by 250% over the next two decades; hence, based on the projected air traffic growth, aviation weather capabilities are essential to increase the safety parameters of airborne aircraft (Craig, 2012). Human factors are dynamic in nature and change when new technologies emerge because innovations change the flow of cockpit operations and management and flight procedures (Craig, 2012). Most of the scientific data on human factors stem from air transport and military aviation sectors (Craig, 2012); however, this scientific data is applicable to general aviation as TAA are outfitted with the exact same capabilities as commercial and military aircraft (Whitehurst & Rantz, 2012).

The FAA's Human Research Division has oversight over the Flight Deck Human Factor Research Program, which is the one regulatory process that provides technical expertise to the entire aviation domain; however, as technological capabilities become more sophisticated the FAA will struggle to develop the human capital to support future requests (Craig, 2012; Downer, 2012). The Flight Deck Human Factor Research Program provides the industry with support by identifying research requirements, executing research efforts, disseminating the research data to sponsors, and

engaging in feedback, in which Craig refers to this as
from research to reality. Aviation experts, scholars,
and industry leaders recommended the FAA
proliferate efforts to make human factors engineering
an essential capability when designing flight decks
and for flight deck design teams to possess adequate
human factors expertise (FAA, 2013) to reduce
design errors that impact human factors.

Business practices in the aviation industry
include technical, human, and organizational
applications, which are integral variables of the socio-
technical systems of system (Craig, 2012). The SMS
practices that provide organizational oversight to
detect hazards at all levels of the socio-technical
construct under the organization's strategic
management functions are capable of identifying
emerging human factors problems (Craig, 2012;
Ropp, 2009). Safety management system is a multi-
disciplinary concept in which human factors
awareness is an integral part of the education and
training program (Ropp, 2012). Human factors
awareness is a major concern in general aviation due
to integration of glass cockpits, because existing flight
training programs are mostly aligned to non-TAA

(Lindo et al., 2012). The Future Aviation Advisory Committee recommended the implementation of additional technologies to improve reliable, efficiency, and situation awareness; however, the committee caveated that aviation stakeholders have to undertake and reduce the intrinsic systemic hazards of implementing technological innovations (USDOT, 2012).

Aviation technology developers have to keep the human centered design model in the forefront of all developments to lessen the confusion and anxiety for pilots (Craig, 2012). Human factors problems in the aviation also caused by poor oversight because managers have the ability to regulate systems (Smith, 2011) and use predictive analysis to identify potential problems (Halladay, 2013). Leaders in the general aviation organizations are responsible for the directing and managing the systems within because the essence of SMS is the orchestration of complex systems to alleviate hazards, injuries, or mistakes (Smith, 2011).

Operating TAA is different from operating conventional aircraft because pilots are required to adjust to managing information rather than focusing

on traditional tasks when operating non-TAA
(Halleran & Wiggins, 2010). General aviation pilots
struggle to transition from conventional aircraft to TAA
and develop the higher order thinking skills to attain
proficiency in aircraft with glass cockpits (GAO, 2012;
Lindo et al., 2012). Di Renzo (2010) postulated
modern aviation innovations are pervasive and the
current integration pace of new technologies in the
aviation industry prevents the FAA from
comprehending the magnitude of the challenges. The
impact of technological deterministic thinking on
general aviation is what Batteau (2010) refers to as
danger to the periphery due to the effects caused to
outlying processes and components within the system
of systems. Researchers are advocating safeguarding
the socio-technical system by reducing the negative
effects on human factors, safety, training, regulations,
and aviation maintenance (Parasuraman, & Manzey,
2010; Smith, 2011).

According to NTSB (2010), another problem
plaguing the general aviation sector is the integration
of poorly designed technologies that adversely affect
human factors. Minimizing the role of humans in
automated systems and preventing humans from

solving complex problems in the process have increased the adversity of automation because high levels of automation diminish human interaction, which results in degradation of system knowledge and basic skills (Cassens et al., 2011; GAO, 2012). A recent report by the FAA declared that most general aviation pilots lack the comprehensive knowledge to operate highly automated aircraft without undergoing rigorous training (NTSB, 2010). New aircraft designs and technologies lack intuitiveness, which is confusing to general aviation pilots; thus, requiring new flight training programs that instruct general aviation pilots on the diverging differences between TAA and non-TAA (Di Renzo, 2010). Mitchell et al. (2009) asserted "particular research focus was on training, that is, the conversion of pilots from the use of gauges to operating automated systems, safety, design aspects, situational awareness, the role and responsibilities of pilots, workload, levels of skill, and operational aspects" (p. 14). The preponderance of studies that Mitchell et al. (2009) referenced pertains to the commercial pilots and not general aviation pilots; however, the general aviation community could leverage the lessons learned to reduce the negative aspects of implementing technologies.

Figure 14 - Instrument Training Flights; Photo courtesy of Matthew Piatt.

Improving Flight Training to Address New Technologies.

Whitehurst and Rantz (2012) conducted a feasibility study to determine the degradation of piloting skills when transitioning between TAA and conventional aircraft. One of the reasons for conducting the study was the lack of empirical research on pilots transitioning between TAA and conventional aircraft (Whitehurst & Rantz, 2012).

Currently, student pilots are introduced to TAA from

the onset of flight training, which prevents the pilots

from learning to fly conventional aircraft. Whitehurst

and Rantz stressed the significance to address the

importance of learning to fly conventional aircraft due

to the large number conventional aircraft in the

general aviation inventory. Teaching student pilots to

fly TAA is beneficial given that the same automated

capabilities in TAA are used in commercial airliners;

however, the problem arises when the student pilots

attempt to fly aircraft with analog instrumentation,

which the pilots have not been trained to fly

(Whitehurst & Rantz, 2012). Whitehurst and Rantz

stressed that the Federal Aviation Regulations,

specifically Title 14 Part 61.31 does not address

training requirements for pilots transitioning between

TAA and conventional aircraft, which increases the

probability of accidents and incidents due to

expanding number of TAA in general aviation.

Aviation researchers postulated that pilots training in

TAA are capable of the passing the minimum test

standards; however, the training paradigms do not

provide enough in-depth lessons to adequately

prepare pilots to operate advanced technologies (Whitehurst & Rantz, 2012). Whitehurst and Rantz concluded a full-scale study was necessary to determine the differences in operating TAA and conventional aircraft.

The continuous integration of technological innovations influences the general aviation sector by requiring new flight training requirements such as teaching pilots to manage flight information rather than focusing on the development of technical skills (Robertson, 2010). The integration of TAA in general aviation obliges the aviation community to add training requirements to teach pilots to be proficient at using automation and higher order thinking skills to maintain adequate levels of situational awareness, mental alertness (Robertson, 2010), and better decision-making skills (Halleran & Wiggins, 2010). Some general aviation organizations have implemented FITS to overcome the training deficiency; however, there are no regulations mandating recurring training requirements for FITS, especially for pilots that drop flight currency. Other studies concluded with similar results as Robertson's (2010) experiment, which recommended teaching pilots to become flight managers since the automated capabilities require pilots to transition from using psychomotor skills to managing advanced avionics suites and sensors.

Some researchers recommended major improvements in flight training; however, the FAA has not made any regulatory changes directing new training paradigms (Lindo et al., 2012). Researchers recommended the following suggestions to reduce the adverse effects of integrating TAA in general aviation, including: (a) scenario-based training, (b) single-pilot resource management, (c) aeronautical decision-making, (d) automation management, and (e) task management as training initiatives to improve general aviation flight training (Cassens et al., 2011; Lindo et al., 2012; NTSB, 2010). Members of the NTSB (2010) noted the FAA had not updated the knowledge tests to include testing information on glass cockpits nor have specified TAA training requirements been included in the practical test standards. "With the exception of training provided by airplane manufacturers with the purchase of a new aircraft, pilots must currently seek out and obtain equipment-specific glass cockpit training on their own" (NTSB,

2010, p. vii). This highlights the training gaps in general aviation and the disjointed and onerous efforts between the FAA, NTSB, academia, and GAJSC. The FAA, NTSB, and GAJSC are advocating for FITS as the leading training model in general aviation (NTSB, 2010). Geiselman, Johnson, and Buck (2013) postulated that training is only one of the critical elements to address the integration of advanced capabilities. As the next generation of technologies are integrated into aircraft, designers and industry leaders have to develop holistic training programs to prevent negative effects like automation surprise (Geiselman, Johnson, and Buck, 2013).

The FAA struggles to maintain a proactive approach to provide government oversight on regulating pilots transitioning between conventional aircraft and TAA and vice versa due to a shortage of aviation safety inspectors (Downer, 2012; Lindo et al., 2012, USDOT, 2012). Conventional aircraft are airplanes equipped with steam gauges and lack the sophistication of TAA (Lindo et al., 2012). General aviation pilots transitioning between TAA and conventional aircraft do not meet the risk threshold for the FAA to investigate this matter due to a personnel shortage (USDOT, 2012). Aviation experts do not consider this negligence. The FAA is in the process of delegating tasks to private companies like aircraft manufacturers (USDOT, 2012). Members of the NTSB (2010) and Lindo et al. (2012) determined thus far that TAA are not improving general aviation safety; consequently, TAA have led to technological complications and exposed regulatory and training gaps.

Now that TAA are becoming increasingly popular, the general aviation sector is experiencing similar accidents as military, corporate, and commercial sectors (Lindo et al, 2012). Pilots in the military, airline transport, and corporate aviation sectors receive more training on a recurring basis than general aviation pilots do; yet, general aviation pilots are operating aircraft with the same levels of automation and technological capabilities (NTSB, 2010). The paucity of standardized flight training further compounds the problem due to general aviation being less regulated than commercial and military aviation (Lindo et al., 2012). Disjointed efforts between academia, the government, general aviation pilots, the aviation industry, and other aviation organizations have resulted in a disconnection between contemporary training methodologies (Pourdehnad & Smith, 2012). With the implementation of new aircraft with different technologies and flight handling characteristics, the abovementioned stakeholders need to place more emphasis on determining the most effective training rather than industry growth (Cassens et al., 2011).

Current flight training paradigms focus on teaching pilots to manipulate the automation; however, the same training models do not include objectives to teach the pilots to manage the automation during emergencies (Martins, Martins, Soares, & Augusto, 2013). A misconception in the aviation industry is "while investment in automation increases, less investment is needed in human skill" (Martins et al., 2013, p. 155). A consequence of introducing new automated capabilities in TAA is the demand for additional training, skills, and mitigating the human factor issues (Martins et al., 2013). Human factors subject matter experts deem it necessary for managers of aviation companies and aircraft manufacturers to identify hazards or predictable errors associated with the implementation of new technologies (Martins et al., 2013). Presently, regulations do not mandate aviation companies or aircraft manufacturers identify predictable errors that lead to human error, which is compounded by the lack of a criteria or system to detect problematic errors, which requires an immediate regulatory change to address the oversight (Martins et al., 2013). Aviation industry experts postulated the need for aviation organizations to include human factors specialists in the design selection of aviation technologies to reduce human factors issues (FAA, 2013). One area of concern is flight deck standardization because the lack of uniformity leads to pilot errors, negatively affects learning, and increases flight training (FAA, 3013). However, aviation researchers are aware that revamping all aviation technologies is not feasible and may impede technological advancements (FAA, 2013).

To reduce the adverse effects of technological developments, the general aviation community needs a comprehensive approach consisting of new training requirements that are regulated and mandatory (Cassens et al., 2011; Halleran & Wiggins, 2010; Manzey, 2012). The lack of regulatory oversight

perpetuates the problems since training and the
implementation of regulations lag behind the
integration of new technologies, which puts general
aviation pilots at a greater risk (Downer, 2012; Lindo
et al., 2012). The impacts of integrating new aviation
technologies in general aviation are unknown; hence,
the need for further exploration to understand and
overcome this phenomenon.

*Figure 15 - Garmin G1000 panel of Diamond DA-42 Twin Star,
N49494, based @ KBFI - Seattle, WA; photo courtesy of
Matthew Piatt.*

Capitalizing on Innovations by Reducing the Negative Influences.

Placing more emphases and efforts on training, management, technology selection, and human factors deficiencies will allow the aviation industry to leverage the benefits of the technology (Halleran & Wiggins, 2010). A benefit of automation is to improve safety; however, the integration of TAA in general aviation introduced new human factors, cognitive, and flight training problems (Cassens et al., 2011; Lindo et al., 2012). Feary et al. (2010) asserted extensive usage of automation has led to more accidents in the aviation community. To reduce the adverse effects, "more robust methods and tools are needed for predicting and evaluating automation interaction early in the design process" (Feary et al., 2010, p. 36). One methodology to pre-test the risks and hazards associated with new technologies is to use experienced pilots to establish a baseline for appropriate pilot behavior during a regulated experiment (Feary et al., 2010). The behavioral responses of experienced pilots will be compared to pilots with lesser experience to develop initiatives to reduce the hazards associated with new technologies (Feary et al., 2010). This behavior modeling is a practical method to safeguard against the increasing levels of complexities of emerging technologies (Feary et al., 2010). The advantage of using behavior modeling is to target problems caused by new technologies (Feary et al., 2010).

The FAA is directing the implementation of the Next Generation Air Transportation System (NextGen) to improve convenience, dependability, and safety (Berry & Pace, 2011). According to Berry and Pace (2011), one of the major benefits of NextGen is it will modernized the national airspace system to handle the increased air traffic volume and to ameliorate airspace safety. Researchers avowed the integration of new technologies create unforeseen problems (Pritchett, 2009); thus, making it essential for managers to mandate the use of predictive analysis to reduce unexpected problems (Halladay, 2013). Predictive analysis forecasts the behaviors of air traffic controllers, dispatchers, and pilots using NextGen or aviation technologies and provides outcomes, in which the statistical data enable aviation leaders to make informed decisions to reduce unanticipated problems (Berry & Pace, 2011; Halladay, 2013). The NextGen will influence the performance of general aviation pilots, air traffic controllers, and dispatchers (Berry & Pace, 2011); thus requiring aviation leaders to understand the impact of NextGen on the current socio-technical system by synthesizing the effects on each component of the system (Smith, 2011). Reducing the negative aspects of NextGen is critical for flight safety and improving the efficiency of air traffic new technologies (Lee, Jeon, & Choi, 2012).

Safety management system (SMS) is an organizational approach to enhance safety awareness by developing a solid foundation through organizational cultural that encompasses the fundamental practices of reducing aviation accidents (Flouris & Yilmaz, 2009; Lu et al., 2011). A SMS consists of three sub-systems (a) safety organization and tasks, (b) safety procedures, and (c) discharging of corrective actions (Naveh, Katz-Navon, & Stern, 2011). The airline transport companies are implementing SMS to enhance its cultural approach to avert aviation accidents and incidents by changing existing approaches to ensure safety compliance to ensure every individual within the organization is empowered to notify organizational managers or safety officials of any potential safety violations (Flouris & Yilmaz, 2009).

The FAA mandated all airline carriers implement a SMS (Lu et al., 2011); however, there was no regulatory guidance given to the general aviation sector regarding SMS (GAO, 2012). A strategic goal of the FAA is to decrease general aviation accidents by 1% per year to 2018 (GAO, 2012); yet, the FAA is struggling to provide regulatory oversight to achieve the annual 1% reduction. Some researchers argued that general aviation organizations are not postured to leverage a SMS due to its non-traditional hierarchies because in general aviation operations are typically single-piloted, unlike commercial aviation operations (Garibay & Young, 2013). Most general aviation operations do not have the oversight or checks and balances like commercial operations (Garibay & Young, 2013). Garibay and Young declared the implementation of SMS in general aviation would be impeded by the lack of motivation and incentives; hence, creating an organizational barrier. Safety management systems were implemented as a systematic approach to improve organizational safety for commercial airlines to exceeding compliance standards; however, the general aviation can benefit a SMS (Garibay & Young, 2013).

A critical phase of SMS that general aviation can use immediately is identifying areas requiring change and taking the appropriate actions to bring awareness to organizational leaders. Naveh et al. (2011) declared this phase as a continuous improvement stage, which applies to general aviation as a business practice. Flouris and Yilmaz (2009) postulated, "the efficient and effective management of any aviation organization, regardless of the nature of its functions or its size, requires the management of basic and traditional business processes: financing, budgeting, communicating, allocating resources and so forth" (p. 12). Due to the increasing significance of safety, organizations have added safety management as a core business practice; primarily most institutions refer to this business approach as safety risk management (Flouris & Yilmaz, 2009). Based on Flouris' and Yilmaz' (2009) supposition that safety risk management (SMR) is a business practice, then general aviation organization could pursue SMS as a methodology to identify safety risks and hazards. Leveraging SMS in general aviation will provide the foundation to evaluate the impacts of new technologies, in which the SMR approach will identify vulnerabilities associated with each phase of operation (Flouris & Yilmaz, 2009). Implementing a SMS appears to be advantageous; however, some organizational leaders and researchers are apprehensive of the cost associated with establishing a SMS (Duncan, 2013). Opposing leaders view SMS as a liability to an already austere financially constrained environment (Duncan, 2013).

Scientific literature indicated that automation could improve the proficiency, performance, reliability, and accuracy of operators (Hughes et al., 2009); however, additional research is necessary to understand the adverse effects of technological developments. Pritchett (2009) acknowledged, "recent experience has shown that instead of eliminating error, automation accentuates some weaknesses in aviation operations and generates new types of error arising from problematic human-automation interaction" (p. 83). There have been some successes with integrating new technology; however, more research is required to reduce the negative effects of aviation innovations. General aviation pilots need to understand any inherent risk connected with technologies to prevent the influences from affecting aviation safety to include aviation maintenance.

Management Practices in General Aviation.

The GAO (2012) categorized 78% of general aviation activities coincide with four areas of operations: (a) personal, (b) business, (c) corporate, and (d) flight instruction. For the most part, Federal Aviation Regulations Part 91, general operations and flight rules apply to general aviation (GAO, 2012). There appears to be theoretical disconnect between regulatory authorities and the general aviation community due to the lack of learning from pragmatic errors in a high-risk domain (Catino & Patriotta, 2013) such as the airline transport industry. Researchers indicated that re-evaluating previous mistakes prevents similar mistakes by affording organizations the ability to improve safety, compliance, and assurance aspects to break the chain of mistakes

(Bergeon & Hensley, 2009; Catino & Patriotta, 2013).

An employee of a regulatory authority declared the

process to prevent accidents as the Swiss Cheese

Model, which entails developing mechanisms such as

laws, regulations, and processes to avert aviation

safety accidents (Bergeon & Hensley, 2009). The

impacts of innovations on general aviation require

practitioners, managers, and safety specialists to

develop an in-depth, theoretical understanding to

leverage resources like SMS, learning theories, and

predictive risk mitigation analysis (PRIMA) to

heightened safety awareness in general aviation

(Bergeon & Hensley, 2009; Catino & Patriotta, 2013).

The urgency of heightened safety awareness is

because 92% of general aviation accidents in 2011

were fatal (GAO, 2012). As of 2011, there were

580,800 active pilots, in which the FAA invested 203

million dollars to reduce the number of fatal general

aviation accidents (GAO, 2012).

According to Freiwald, Lenz-Anderson, and

Baker (2013) a large number of certified flight

instructors in the United States have 12 months or

less of aviation experience. Most of the certified flight

instructors in the United States provide flight instruction to attain flight hours to meet a minimum flight hour requirement to become a regional airline pilot (Freiwald, Lenz-Anderson, & Baker, 2013). The lack of experienced certified flight instructors in the United States affects the quality of flight instruction, especially with the integration of TAA in general aviation. Because of high turnover, the general aviation sector relies on newly qualified certified flight instructors to provide flight training; consequently, some researchers are predicting an increase in general aviation accidents (Freiwald, Lenz-Anderson, & Baker, 2013) because of the increasing complexities associated with aviation automation (King, 2011). King avowed system knowledge expertise, use of scenario-based instruction, and enhancing aeronautical decision-making are factors that could improve general aviation safety.

Catino and Patriotta (2013) asserted learning theories address cognition, sensation, and values to address the chasms in learning. This belief aligns to assertions made by practitioners that emphasize retooling pilots' cognitive abilities and the general aviation's cultural assumption on operating TAA

(Halleran & Wiggin, 2011). The general aviation sector needs theoretical and scientific research on predictive risk mitigation analysis and learning theories coupled with managerial oversight processes to change the culture of leveraging business practices to improve organizational safety (Bergeon & Hensley, 2009; Catino & Patriotta, 2013; GAO, 2012). A common and risky practice of aviation organizations like the Future Aviation Advisory Committee is to expedite the development of aviation technologies (USDOT, 2011). From a practitioner viewpoint, dynamic research and strategies between government and private organizations will ensure the United States remains an international competitor in producing technologies (USDOT, 2011); however, researchers have failed to address the regulatory efforts to evaluate the rulemakings for emerging technologies (Downer, 2012).

In 2011, members of the FAAC recommended the aviation industry and FAA allocate adequate resources to collect and analyze data from various safety programs (USDOT, 2011). Pourdehnad and Smith (2012) declared, "the aviation industry has been successful in increasing safety, but it has failed to utilize the teachings from dealing with complex socio-technical systems and apply it to the business of aviation" (p. 84). This is indicative of an industry that has failed to learn from previous mistakes and lack of regulations regarding integration of glass cockpits in general aviation (Lindo et al., 2012).

Scientific, fiscal, material, and social limitations affect government organizations, to include the FAA (Boyne & Meier, 2009). These limitations prevent public institutions like the FAA to maintain effectiveness during fiscal constraint periods (Boyne & Meier, 2009). Fiscal limitations prevented the FAA from providing effective oversight on technological innovations (Downer, 2012). The FAA is a diverse institution that provides numerous services to entire the aviation field (Boyne & Meier, 2009). A common belief across aviation is that implementing technologies increases adeptness, trustworthiness, and ameliorates safety (Jackson, 2010). Jackson argued innovations become widespread capabilities followed by regulatory decisions on emerging technologies. The primary objectives of the regulatory authority are safety compliance and oversight and to ensure the aviation industry remains a vital economic enabler; (Jackson, 2010). Scholars noted long-term effects to fiscal, social, political, or material implications are indeterminate and have negative consequences to governmental functions (Boyne & Meier, 2009) such as the FAA.

The paucity of government oversight on technological innovations is due to the increasing complexities of modern aircraft, limited budget, and human capital, and a shortage of subject matter experts to make decisions on the technologies the FAA certifies (Downer, 2010). Even though these problems are centric to the aviation industry, general aviation will be impacted due prioritization when compared to commercial and military aviation due to staffing, resourcing, and oversight constraints within the FAA (USDOT, 2012). During an International Air Safety Seminar, a senior FAA official postulated that safe designs, operations, and maintenance support is the responsibility of the manufacturing company (Downer, 2012). In a report to Congress, the GAO classified the FAA's staffing and personnel allocation process as ineffective, which prevents the regulatory agency from providing essential oversight responsibilities (GAO, 2012). As the United States' sole regulatory authoritative, the general aviation community should be concerned with the overall efficacy of the FAA because the shortage of key personnel creates vulnerabilities (GAO, 2012). The insidious beliefs of implementing technologies to reduce general aviation accidents have not yielded plausible results; hence, forcing industry experts to develop new strategies to reduce general aviation accidents (GAO, 2012). The ability to develop a risk-based approach to reduce general aviation accidents is not feasible because the FAA is delegating essential certification responsibilities to aircraft manufacturers, technology developers, and private organizations (USDOT, 2012).

The integration of TAA in general aviation is changing the organizational culture of the general aviation. Donahue and O'Leary (2011) equated an organization's culture to individual's personality because every entity has a different culture. Technological deterministic thinking affects society and culture (Chandler, 2012) because culture is the collective norms and understanding established by employees (Donahue & O'Leary, 2011). Organizations that effectively implemented change used action vehicles, which are mechanisms transformed concepts into procedures, structures, and processes (Donahue & O'Leary, 2011). General aviation could benefit from action vehicles by solidifying practices, standard operating procedures, and flight training to reduce the influences of technological developments (Donahue & O'Leary, 2011; Lindo et al., 2012). Researchers emphasized the significance of integrating changes into the fabric of the organization to enhance the momentum of the changes (Donahue & O'Leary, 2011). The general aviation community is undergoing changes due to the integration of technological developments, which is affecting safety because general aviation lacks the managerial oversight and consists mostly of pilots conducting single piloted operations (Garibay & Young, 2013).

Many organizations struggle to implement effective safety programs because the programs fail to provide feedback or data for continuous improvements (Blair & O'Toole, 2010). Some researchers argued that uncertainty is inherently part of the innovation process due to the lack of scientific research on the factors that drive the uncertainty

(Jalonen, 2012). Historically, general aviation accident rates have always been higher than commercial aviation due to the disparity in training, innovations, pilot experience levels, and single piloted operations (Garibay & Young, 2013). Nevertheless, the proliferation of TAA in general aviation affects the abovementioned areas because the aviation industry lacks government oversight due to staffing shortages (GAO, 2012), the uncertainty of new technologies (Jalonen, 2012), and general aviation pilots' limited resources and flight support when compared to the airline pilots (Garibay & Young, 2013). Commercial pilots receive mission and flight support from dispatchers and aviation specialists on the ground for in-flight emergencies or during flight operations (Garibay & Young, 2013), which general aviation pilots do not receive.

Organizational managers are affronted with making effective decisions concerning the implementation of technology (Ganguly, Nilchiani, & Farr; 2010) to include managers at the FAA (USDOT, 2012). In 2010, President Obama announced his Strategy for Innovation that addressed the federal government as a leader in innovation through investing in pragmatic research, human resources, and infrastructure (Farrill & Kalil, 2010). Farrill and Kalil posited that the nation's technological infrastructure is declining due to fiscal constraints and political and economic scrutiny. Even though the United States failed to advance its technological infrastructure, the country retained its international prowess in producing technological innovations (Farrill & Kalil, 2010). The lack of technological infrastructure has affected the aviation industry, which is indicative of the FAA lacking the human capital to maintain technological and managerial oversight demands of the aviation industry (Downer, 2012).

Some senior executives avowed effective business management in the aviation industry consists of overseeing numerous business processes, to include SMS, because safety is a fundamental business function (McNeely, 2012). O'Toole and Nalbone (2011) emphasized senior managers have to ensure resources are available to support safety initiatives to achieve fewer safety incidents and maintain a healthy safety culture. Research studies indicated the most important factor in the SMS approach is support from senior managers (O'Toole &

Nalbone, 2011). The complex socio-technical environment confronts managers in the aviation industry because failures in such systems are singular in nature; however, the outcomes affect the entire infrastructure (McNeely, 2012) due to the system of systems construct (Craig, 2012). Even though SMS is a holistic approach to improve and manage organizational safety, scientific literature on SMS does not indicate that general aviation organizations are effectively employing SMS (Ropp, 2012). The FAA directed the implementation of SMS in general aviation to teach pilots to leverage all applicable resources to ensure a safe flight because personal operations resulted in the highest number of general aviation accidents (GAO, 2012). Due to the high number of general aviation accidents, a report to a congressional committee indicated that the FAA is not directing enough efforts on general aviation (GAO, 2012). The benefit of SMS is the ability to identify emerging hazardous (Ropp, 2012) such as new threats from innovations.

The current administration's inability to slow the Unites States' technological decline requires a meticulous plan to avoid an influx of disruptive

technologies (Bendrath & Mueller, 2011). Disruptive
technologies are innovations that destabilize current
social constructs, norms, markets, and institutional
practices (Bendrath & Mueller, 2011). According to
Bendrath's and Mueller's (2011) definition of
disruptive technologies, the integration of TAA in
general aviation is disruptive and unsafe due to the
imposed effects on existing practices in general
aviation. Technological innovations are disruptive
when the breakthroughs cause adverse effects on
societal and institutional practices because
innovations require synchronization and integration
into the existing organizational culture (Bendrath &
Mueller, 2011). Integration conflicts develop from
unsynchronized efforts to implement technologies,
which require domain or institutional changes to avoid
technologies from disturbing existing practices, or
political implications (Bendrath & Mueller, 2011).

Bendrath and Mueller (2011) alluded the
implementation of technologies has consequences on
public policies, regulations, and laws. The integration
of TAA in general aviation is an example of an
innovation that affected institutional practices in
general aviation; yet, the FAA has not implemented

any regulatory changes to address the integration of glass cockpits in general aviation (Cassens et al., 2011; Lindo et al., 2012). The FAA recommended FITS as an approved flight-training paradigm to overcome the challenges of TAA (Cassens et al, 2011). Currently, FITS is not mandatory nor do existing flight training syllabi provide effective instructions on TAA (Cassens et al., 2011; Lindo et al., 2012). The existence of TAA in general aviation accompanied by the lack of regulations to address the innovative change will grow increasingly troubling as general aviation aircraft become more complicated without regulatory changes. Researchers believe aircraft manufacturers and certified flight instructors are determining the training requirements for TAA ,because FITS training is not a mandatory requirement (Lindo et al., 2012).

The absence of stricter guidance from the FAA may result in general aviation pilots receiving substandard training when transitioning to TAA (Lindo et al., 2012). Some researchers postulate that civil aviation are challenged by regulatory and safety goals because of the dynamism of the civil aviation system (Layton, 2012). Policy makers demand guidance on the adverse effects of new technologies to characterize associated risks that may cause societal disruptions (Carlsen et al., 2010). Policy guidance is necessary to understand the negative implications on critical infrastructure (Carlsen et al., 2010) in which the air traffic control system is an essential component of the aviation system (Berry & Pace, 2011). To prevent new technologies from being disruptive, an assessment is required to determine the effects on social decision-making (Carlsen et al., 2010). The results of the assessment will determine if the technology is integrated, regulated, promoted, or restricted (Carlsen et al., 2010).

Figure 16 - Cockpit of Northwest Airlines, DC-9-40 (reg. 9760). Picture taken on the ground at Port Columbus International Airport, Columbus, OH; photo courtesy of Dmitry Denisenkov.

Collaborative efforts by air traffic specialists, aviation managers, airport managers, communication and navigation representatives, pilots, and aviation support organizations are addressing the concerns; yet, the problems still exist (Layton, 2012). The reasons why accidents are continuing to occur are: (a) current methodologies do not discover the root causes of accidents, (b) learning from previous mistakes is an erroneous approach, and (c) failure to capitalize on flight simulator training (Leveson, 2011). Researchers argued systems and organizations are constantly undergoing changes to accomplish short-term objectives and reduce costs (Leveson, 2011). Leveson avowed reliance on technological innovations and digital technologies was prevalent; hence, changing current practices and approaches. Even though scientific research has proven the relationship of cost of safety versus profitability to be indeterminate, the datum only reflects on private organizations and not public organizations (Madsen, 2013).

In some cases of disruptive technologies, politics drive technologies, or emerging technologies will change existing regulations (Bendrath & Mueller, 2011). The integration of TAA in the general aviation sector has not triggered any changes to existing aviation regulations (Lindo et al., 2012). According to O'Connor et al. (2011), government authorities assume that regulatory changes are not necessary due to a decline in the number of general aviation accidents and false positives stemming from lagging indications. Some researchers emphasized the importance of risks and vulnerabilities analyses to determine the effects of technological innovation on society (Carlsen et al., 2010).

Risk analysis exposes the nature and characteristics of the adverse effects of technologies by analyzing the probability and severity of risks through an expected utilization model (Carlsen et al., 2010). Vulnerability analysis is identifying weaknesses in a system that restricts innovations from performing as designed when threats and hazardous are introduced (Carlsen et al., 2010). Due to the risk and vulnerabilities of new technologies, national authorities routinely evaluate the adverse implications on critical infrastructure (Carlsen et al., 2010); the national airspace system is a critical infrastructure (Sharma et al., 2009). The evaluation of critical infrastructures is based on qualitative and practical analysis to prevent technologies from becoming disruptive to socio-technical systems (Craig, 2012; Carlsen et al., 2010). Some researchers indicated that a theoretical framework is necessary to gain an in-depth perspective on how technology changes society (Carlsen et al., 2010). Chou and Zolkiewski (2012) proclaimed that organizational leaders are responsible for observing and controlling the dynamic changes stemming from technological innovations and the interaction between new technologies and existing innovations.

To determine the appropriate framework and role of the federal government, it is necessary to ensure civil aviation remains safe and a vital economic resource. Some critics call for government to relinquish its regulatory responsibilities so aviation stakeholders can guide the architecture to regulate the industry (Farrill & Kalil, 2010). Academia, general aviation, private, and aviation industry leaders are vouching for an amalgamated approach between government and private entities to shape the framework and process for implementing innovations (Farrill & Kalil, 2010). Farrill and Kalil declared the federal government is responsible for developing a framework that cultivates technological innovations to the create jobs and build healthier industries. The Obama Administration is developing the foundation to garner progressive innovations (Farrill & Kalil, 2010), in which technological developments will continue to expand in the civil aviation sector.

The lack of managerial and regulatory underpinnings compels the engagement in cyclical practices analogous to FITS to offset the adverse attributes of innovations in general aviation (Lindo et al., 2012). Technological deterministic assumption is forcing organizations to transform from traditional business practices to rely on competence management processes (Corallo, Lazoi, Margherita, & Scalvenzi, 2010). The widespread emergence of technologies is causing disruptions driving changes in markets, product developments, and management practices (Corallo et al., 2010). Traditional practices for managing human resources are undergoing transformation due to human task and competencies required to work in technological driven and based organizations. For instance, certified flight instructors are required to instruct general aviation pilots on operating TAA; thus, requiring to an amendment to teaching paradigms to provide period of instructions on FITS and aeronautical decision-making (Cassens et al., 2011). Competence is the fundamental aspect that relate to performing specific tasks by knowledge, skills, and behavior (Corrallo, 2010). An impact of technological determinism on the general aviation sector is transforming existing competencies to account for the integration of technological innovations and how the innovations affect standing practices and procedures (Corrallo, 2010). Competence management is a desirable capability for general aviation organizations to ensure businesses have the human capital and talent to overcome constant technological changes (Corrallo, 2010).

Organizational change is effective leadership that positively cultivates transformations accompanied by inventive thinking to achieve strategic objectives (Attwood, Mora, & Kaplan, 2010). To understand the organizational learning and organizational change discourse it is best to view general aviation sector as a separate entity. The FAA is a public organization that manages regulatory matters, which provides a service to the aviation industry; therefore, the general aviation sector is a customer of the FAA (Mikel-Brumfield, 2009). Mikel-Brumfield asserted federal organizations such as the FAA should use a best practice approach, which is not the case with integrating TAA in the general aviation. Effective leaders are able to foster a collaborative environment to establish common objectives by leveraging resources, intellectual and human capital to create an environment of learning, adapting, and organizing (Attwood et al., 2010).

The existing aviation safety construct is a bi-level approach with the first layer consisting of codified regulations, procedures to address technological changes, and risk knowledge management (Jackson, 2010). The second layer includes regulatory oversight that involves the incorporating or changing of existing regulations, policies, and risk management practices (Jackson, 2010). Regulations are legal policies that provide specific directions to stakeholders managing, directing, or operating within the civil air domain or

national airspace system (Jackson, 2010). Some
researchers believe aviation regulations derive from
catastrophic aviation events or adverse situations
(Jackson, 2010); yet, the societal, institutional, and
safety implications from integrating TAA in general
aviation haves not yielded any regulatory changes
(Lindo et al., 2012; Misa, 2009). Aviation managers in
the general aviation sector have to transition from
traditional managerial roles to act as enablers to
pursue the collaboration and coordination to foster the
oversight to manage the sector's safety interests
(Denning, 2011). The FAA is responsible for
coordinating directives through Title 14 of the Code of
Federal Regulations (Jackson, 2010); however, there
is a chasm between the general aviation sector and
the FAA on developing new regulations. The FAA is
currently undergoing fiscal constraints like all federal
organizations; therefore, the private sector will have a
lead role in research and development (Farrill & Kalil,
2010).

According to Downer (2010), the FAA needs a
major organizational transformation to improve its
leadership capacity, regulatory oversight, partnerships
and outreach, and a committed effort to educate its

workforce. Due to technological developments being more complex, the FAA does not have the technical expertise to evaluate new technologies (Downer, 2010). This highlights the FAA's inability to provide oversight and governance for emerging technologies (Lindo et al., 2012). Dumas et al. (2010) stressed the significance of organizational leadership possessing the aptitude to implement institutional practices, processes, procedures, and systems that enhances organizational learning. Limited financial resources accompanied by a diminishing technical capacity impede the FAA from eliminating hazards and vulnerabilities (Downer, 2010). This includes the ability to implement effective business practices and processes to integrate new technologies. Such degradations and limitations led to leveraging programs and partnerships such as FITS and AvSP, which are not mandated by the FAA and serve as initiatives to improve aviation safety (Downer, 2010; Sharma et al., 2009). The continuous integration of technological developments in aviation has exceeded current policies and the FAA's ability to provide effective oversight (Lindo et al., 2012; Yeh et al., 2012). Another factor affecting managerial oversight is regulatory authorities at the national and international levels have different guidelines, strategies, and interpretations of regulations (Layton, 2012).

Researchers regularly documented that technological innovations affects politics, the populace, and cultures due to technologically driven organizations focusing on capitalistic gains rather than the negative effects that impact policy, normative and cognitive effects, and institutional influence (Misa, 2009). The gap between the general aviation sector and the integration of innovations exist due to stakeholders not fully understanding that solutions in the commercial sector are not always applicable to the general aviation sector (Black & Chimka, 2011).

Figure 17 - Cockpit of an Aeroflot Il-96-300; photo courtesy of Dmitry Petrov.

Summary

Safety experts at the NTSB (2012) postulated the integration of TAA into general aviation did not improve safety; in fact, the automated aircraft caused unintended consequences such as a higher fatality rate in aircraft with glass cockpits. Additionally, the integration of TAA in general aviation called attention to outdated flight training paradigms, human factors concerns, and the lack of managerial oversight (Layton, 2012). As noted by human factors scholars, the integration of new technologies requires a complete evaluation to determine the impact on the aviation socio-technical systems (Lee, Jeon, & Choi, 2012; Smith, 2011; Parasuraman & Manzey, 2010). Human factors experts suggested that general aviation pilots undergo FITS, aeronautical decision-making, and single pilot resource management training to become proficient at operating TAA (Lindo et al., 2012). Technologically advanced aircraft are outfitted with advanced automated capabilities that are intended to reduce the workload and mental burdens associated with flying aircraft; however, to capitalize on the automation requires general aviation

pilots to undergo FITS (Lindo et al., 2012). The

literature on the required training indicated the

general aviation sector needs additional regulations to

govern TAA (Lindo et al., 2012). Technology scholars

avowed that leaders in the aviation industry failed to

provide adequate oversight and strategies to manage

the automation and emerging technologies

(Vanderburg, 2012).

The literature exposed the social and cultural

implications from integrating TAA in general aviation,

which is indicative of technological deterministic

thinking, as evident by the unintended consequences

and disregards for the existing network of systems

(Chandler, 2012; Lee, Jeon, & Choi, 2012). The

failure to understand technological deterministic views

prevents aviation stakeholders from capitalizing on

the positive attributes of the technologies while

minimizing the negative social and cultural influences

(Chandler, 2012; NTSB 2012; Parasuraman &

Manzey, 2010). The literature review revealed that

technological deterministic beliefs affected social,

political, cultural, and educational constructs;

therefore, to reduce the impact requires managers to

assess the changes holistically and implement

corrective actions (Misa, 2009; Vanderburg, 2012).

CHAPTER III
METHODS AND PROCEDURES

In 2011, 92% of all aviation fatalities involved general aviation pilots (GAO, 2012; NTSB, 2010). The integration of TAA in general aviation was necessary to improve safety (FAA, 2013; Garibay & Young, 2013; Robertson, 2010). The integration of TAA in general aviation resulted in shortfalls in safety practices, flight training paradigms, regulatory, cockpit designs and equipment, and operational discrepancies, as well as government oversight concerns. In order to capitalize on the technological advantages of TAA, pilots must undergo additional training to learn to operate new-generation innovations (Garibay & Young, 2013). Researchers, general aviation pilots, and aviation industry leaders need to understand the impact of integrating TAA in general aviation.

A significant problem with integrating TAA in general aviation is the lack of a systematic approach for integrating TAA that addresses impacts on current system designs, processes, and regulations (Garibay & Young, 2013; Pritchett, 2009; Robertson, 2010). The general aviation sector needs a systematic process to integrate technologies without introducing new hazards and safety concerns (Di Renzo, 2010). Without a systematic process, the general aviation sector lacks the ability to identify hazards and implement corrective measures to eliminate unintended consequences such as aviation accident fatalities, improper training, and the lack of regulatory oversight (Cassens, Young, & Greenan, 2011; McCracken, 2011; NTSB, 2012). Robertson's study determined the automated capabilities needed for general aviation aircraft as an effort to target pilot error because 80% of general aviation accidents were due to pilot error. Robertson recommended additional research on general aviation pilots' trust and rational utilization of automation. In another study, McCracken's (2011) research on general aviation pilots transitioning from conventional aircraft to TAA highlighted that a majority of the pilots received substandard scores. McCracken's study substantiated Robertson's research that general aviation pilots were uncomfortable operating TAA after completing initial orientation training. The sentence should read: McCracken recommended research on pilots transitioning into the TAA to identify changes to improve the quality of flight training. Garibay's and Young's (2013) study highlighted the importance of using technologies to improve aviation safety by comparing general aviation and commercial aviation operational procedures. Garibay and Young postulated that TAA improve situation awareness and general aviation pilots need to embrace new technologies. Garibay and Young recommended follow-on research on the efficacy of recurring flight training, improved safety initiatives, and the use of technology to reduce aviation accidents in general aviation. These studies highlighted increased accident fatalities, improper training, and inadequate regulatory oversight due to the lack of a systematic approach to highlight the impacts.

The purpose of this qualitative single descriptive case study was to explore the experiences of general aviation pilots on the integration of TAA and to determine how the automation of TAA can be used to increase aviation safety and improve pilots' decision-making skills. General aviation pilots in Florida, Georgia, and South Carolina were interviewed to provide personal experiences on the impact of integrating TAA in general aviation. The interview data were collected and substantiated along with technical reports, flightlog books, and observation notes. The data collection was essential to understanding the integration of TAA in general aviation due to the unintended consequences such as accidents, fatalities, a training deficiency, and lack of government and regulatory oversight (Cassens, Young, & Greenan, 2011; McCracken, 2011, NTSB, 2012). Technological deterministic thinking and technology cycles will continue perpetuate the hazardous trends in general aviation unless changes are directed to the existing theoretical beliefs on integrating TAA in general aviation. Failing to understand the effects of technology, the levels of automation, and the uncertainty associated with technological needs; increases the probability of more accidents due to automation and information overload (Garibay & Young, 2013; Robertson, 2010).

The qualitative research questions were used to explore the impact of integrating TAA in general aviation and to develop an in-depth understanding of the phenomenon. The intent of the qualitative research questions was to provide the best means to gather data for the described study. Each research question served to gather enriched data on the participants' first-hand experiences with the phenomenon. The following central question and sub-questions were used to explore the impact of TAA on general aviation to align the research purpose with the research problem.

Q1. What are the experiences of general aviation pilots regarding the integration of TAA?

Q2. How can general aviation pilots use the automation of TAA to reduce accidents?

Q3. How can general aviation pilots reduce the effects of technological innovations to capitalize on the positive attributes of TAA?

Q4. What is the impact of TAA on aeronautical decision-making skills of general aviation pilots?

This chapter included a discussion on the research method and design, the validation for determining the sample participants, and the justification for the population. This chapter also included discourse on the material and instruments, the data collection, processing, and analysis procedures, and the assumptions, limitations, delimitations, and ethical assurances.

Research Methods and Design

The purpose of this qualitative single descriptive case study was to explore the experiences of general aviation pilots on the integration of TAA and to determine how the automation of TAA can be used to increase aviation safety and improve pilot decision-making skills. A case study was an effective method to attain contextual reflections of real-world activities to develop a better understanding of how people view routine actions (Miller, 2011; Yin, 2009). A descriptive case study was primarily used to describe a phenomenon by thoroughly scrutinizing the details, propositions and research from the outset of the study (Baxter & Jackson, 2008; Yin, 2009). Existing knowledge of the phenomenon is considered descriptive theory, which is used to established boundaries for the study (Miller, 2011; Yin, 2009). The objective of descriptive case studies was to identify patterns and connections, which will be accomplished with research questions that will answer the propositions (Baxter & Jack, 2008; Yin, 2009).

A case study is primarily used to explore an event, site, or program experienced by individuals through interviews and observations (Giorgi, 2008; Welch, 2011). The reasons for using a case study were: (a) to answer why and how questions, (b) the difficulty to manipulate the data, (c) to collect contextual data that are essential to understanding the phenomenon, and (d) to gain a thorough understanding on the impact of integrating TAA in general aviation (Baxter & Jack, 2008; Yin, 2009). Focusing on this phenomenon was pivotal because there were over 1,600 general aviation accidents per year from 2000 to 2010 (Garibay & Young, 2013). The results of this case study might change the existing theoretical perspective and beliefs on integrating TAA in general aviation.

A case study was appropriate to explore the integration of TAA in general aviation. The number of general aviation fatalities involving TAA doubled the rate of fatalities caused by conventional aircraft and the pilots involved in the accidents are experienced

aviators (NTSB, 2010). The results from this study

highlighted the impacts of integrating TAA in general

aviation, which might lead to theoretical changes and

processes for implementing new technologies in

general aviation. Additionally, this study extended

technology determinism theory and its applicability

and limitations regarding the integration of TAA in

general aviation.

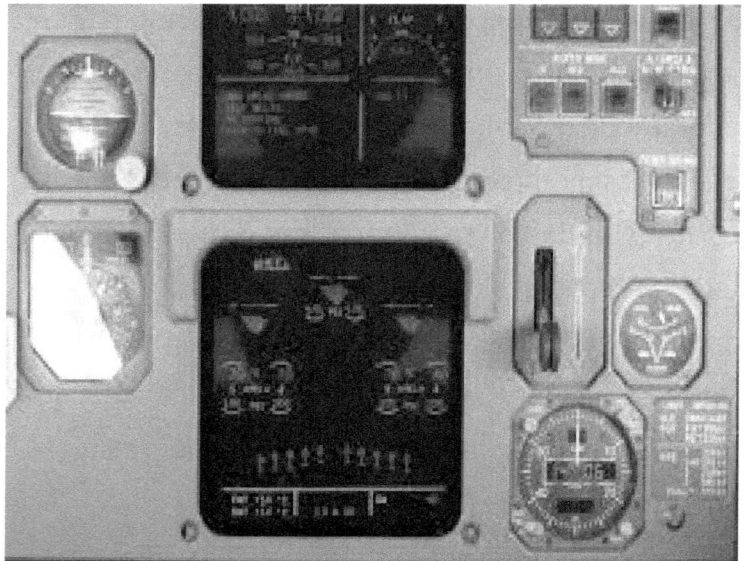

*Figure 18 - The center cathode flight display in the cockpit of an
Airbus A320 family aircraft indicates spoiler deployment and
brake temperature as the aircraft vacates runway 22R at NCE
airport following landing; photo courtesy of Olivier Cleynen.*

This qualitative single case study was important and needed for several reasons. First, there was a scientific knowledge gap between technological determinism and the implementation of TAA in general aviation. Second, the goal of this single case study was to attain a deeper understanding of the impact of integrating TAA in general aviation by using qualitative means. Third, the integration of TAA in general aviation was a phenomenon that requires immediate attention due to increasing number of general aviation fatalities in TAA (NTSB, 2010). Fourth, this study theoretically contributed to technological determinism by providing the aviation industry, academia, aviation organizations, regulatory officials, and technologists with data to understand the impacts of technological innovations.

A Type 2 design, which Yin (2009) classified as a single case study with multiple embedded units of analysis, was used for this study. This case study of determining the impact of TAA in general aviation due to the implication of technological determinism was unique because of the lack of scientific research and interest on the subject (Wong, 2014; Yin, 2009). Additionally, this case study could lead to future research on the integration of TAA, which justified the classification of this as a unique case (Wong, 2014; Yin, 2009).

To gain comprehensive perspectives, this case study explored a single case with three embedded units of analysis or participant groups, which were based on general aviation pilots' flight certifications and qualifications (private pilots, instrument and commercial rated pilots, and certified flight instructors). The benefits of using a single case study with three embedded units of analysis was to develop a deeper understanding on the integration of TAA in general aviation and the impact based on the general aviation pilots' flight qualifications. The embedded units of analysis or participant groups were used collectively to provide comprehensive details of the study. This single case study determined the pilots' experiences on integrating TAA and how to use the automation of TAA to improve aviation safety and pilots' decision-making in general aviation, the first unit of analysis explored general aviation pilots with private pilot certifications, the second unit of analysis explored general aviation pilots with commercial and instrument qualifications, and the third unit of analysis explored general aviation pilots with certified flight instructor qualifications.

It was critical for the participant groups to answer the same interview questions to determine how the integration of TAA affected general aviation pilots. The data collection instruments were semi-structured interviews, which were substantiated with technical reports and studies, flight logbooks, and observation notes. Once data collection and analysis was conducted, the case underwent exploration and comparison of the units of analysis to determine the major themes, patterns, and categories (Rao, 2013; Yin, 2009). The goal was to compare the results of the three units of analysis, and it was evident the research questions answered the propositions (Gao, 2012; Yin, 2009). Finally, after comparing the units of analysis, a case summary was written to highlight the impact of integrating TAA in general aviation.

Another reason to pursue this study was because "technology's impact on societies and individuals highlights an area presently understudied; this need is amplified by reviews of the field and calls for future research" (Cilesiz, 2011, p. 489). A qualitative case study was optimum for exploring and capturing experiences from research participants by semi-structured interviews to explore the phenomenon (Miller, 2011, Yin, 2009). This qualitative single embedded descriptive case study explored the integration of TAA in general aviation and how to use the automation of TAA to improve aviation safety and pilots' decision-making. Exploring multiple units of analysis increased the robustness of the study and reduced the predispositions of a using a singular unit of analysis (Rao, 2013; Yin, 2009). Each unit of analysis independently analyzed and was scrutinized during the data analysis phase (Rao, 2013). The goal of this single case study design was for the case to reveal predictable results that aligned to and supported the propositions (Baxter & Jack, 2008; Yin, 2009).

This case study was based on criteria recommended by Yin (2009), which included the case, the unit of analysis, data sources, and any propositions from theory or literature. A qualitative single descriptive case study with three embedded units of analysis was an appropriate research design to undertake this study for the following reasons.

First, the case for this study focused on the impacts of integrating TAA in general aviation by exploring general aviation pilots (Baxter & Jack, 2008; Yin, 2009). The integration of TAA in general aviation had been problematic as evident by a higher fatality rate and flight training issues (Groff, 2010b). To understand the case holistically, multiple sources of evidence such as semi-structured interviews, technical reports and studies, and observation notes that were used to highlight the case (Baxter & Jack, 2008; Yin, 2009). Thick descriptions were used to identify essential patterns, themes, categories, and clusters to highlight the uniqueness of the case (Mills, Durepos, & Wiebe, 2010). In this qualitative case study, thick descriptions described the impact of integrating TAA in general aviation by highlighting the behaviors and context in a manner comprehensible and meaningful to policy makers, leaders, and technologists (Mills, Durepos, & Wiebe, 2010). Researching the units of analysis allowed the researcher to explore the case.

Second, the units of analysis were general aviation pilots and the data sources, which included semi-structured interviews, technical reports and studies, and observation notes (Baxter & Jack, 2008; Yin, 2009; Zucker, 2001). The unit of analysis was dependent on the research questions because the unit of analysis determined the case within the case study (Baxter & Jack, 2008). Focusing on multiple sources of evidence (Baxter & Jack, 2008; Yin, 2009) was essential to explore the integration of TAA in general aviation. Multiple sources of evidence were optimum for gaining a deeper understanding (Baxter & Jack, 2008; Yin, 2009; Zucker, 2001) on the integration of TAA in general aviation. Using a qualitative single case study approach required a method to explore three embedded units of analysis through data sources.

Third, data sources were semi-structured interviews, technical reports, and observation notes that were taken during the interviews in order to fully understand the impact regarding the integration of TAA in general aviation (Baxter & Jack, 2008; Yin, 2009). Multiple sources of evidence increased the trustworthiness and robustness of the study (Rao, 2013; Yin, 2009).

Fourth, propositions were used to shape and guide the study and direct data collection efforts to

answer the research questions (Baxter & Jack, 2008;
Yin, 2009). Propositions were mechanisms that
brought attention to something to explore in the case
study and where to search for relevant information
(Baxter & Jack, 2008). According to Baxter and Jack
(2008) and Yin (2009), propositions derived from
literature, theories, experiences, and generalizations.
The following were the propositions for this case
study (Baxter & Jack, 2008; Yin, 2009):

a) General aviation pilots value the integration
   of TAA in general aviation
b) General aviation pilots require additional
   training to fly TAA proficiently and safely
c) More governmental oversight and
   coordination is needed for the integration of
   TAA, and
d) General aviation continues to be plagued
   by too many aviation accidents.

These propositions guided the study, shaped
the data collection, and set boundaries for the study
by forming a conceptual framework (Baxter & Jack,
2008; Yin, 2009). Propositions for the study were

limited to prevent becoming overwhelmed with too many propositions (Baxter & Jack, 2008; Yin, 2009). Linking the data to the propositions was accomplished through research questions (Baxter & Jack, 2008).

The interviews were digitally, video recorded with 38 general aviation pilots in Georgia, South Carolina, or Florida (Levasseur, 2011), including the three participants for the pilot study. Data collection instruments were face-to-face interviews, flight logbooks, technical reports, and observations made during interviews (Kvale & Brinkman, 2009; Yin, 2009). While conducting face-to-face interviews, the researcher interviewed 35 general aviation pilots with first-hand knowledge on the integration of TAA in general aviation (Levasseur, 2011; Miller, 2011; Yin, 2009). The interviews were face-to-face to allow the observation and documentation of body language, facial expressions, and other forms of non-verbal activity (Kvale & Brinkman, 2009; Yin, 2009). Additionally, the researcher digitally videotaped the interviews to aid in capturing body language and expressions. The interviews were open-ended questions to explore to the general aviation pilots' experience with TAA. To prevent others from having access to the information, the digitally video

recordings of the interviews were transcribed and

stored on a password-protected computer.

The semi-structured interviews were conducted

at flight schools, flight business offices, flying clubs,

hangars, or airports with TAA operations in Georgia,

South Carolina, and Florida. The study was limited to

the states of Georgia, South Carolina, and Florida to

avoid extensive travel and financial obligations. The

primary researcher is a commercial pilot with first-

hand knowledge of the phenomenon. This made it

imperative to bracket any personal biases or

preconceived notions in order to gather unimpeded

data from the research participants by using an

interview questions guide and asking follow-on and

probing questions during the interviews (Cilesiz,

2011). The goal was to leverage the strength of the

case study design by allowing the personal accounts

of the research participants to authenticate the cases

(Hetherington, 2013; Yin, 2009). This was

accomplished by allowing the research participants to

discuss the impact of integrating TAA in general

aviation without impeding or trying to lead the

participants in a specific direction and adhering to this

approach throughout all phases of the study. Frequent member checking ensured the research participants provided unfettered information (Creswell, 2003; Wofford, Ellington, & Watkins, 2013).

An essential element in qualitative research was the role of the researcher, who was responsible for conducting the interviews, gathering data on people-related experiences with the phenomenon, and interpreting and analyzing the collected data by developing patterns, themes, and details (Cilesiz, 2011; Patton, 2002; Yin, 2009). The interviewer observed the participants' emotions, behaviors, and facial expressions, changes in voice pitch, body language, and noting any other observable aspects during the interviews. After requesting and receiving approval from the research participant, the interviewer videotaped the interviews with digital video equipment for playback to develop a comprehensive understanding of the data via interpretation (Creswell, 2003; Miller, 2011; Schram, 2006). Schram stated the nature of a qualitative investigation was to gain data to develop a deeper and enriched understanding of the phenomenon. During the interviewing, transcribing, and analyzing the data the researcher minimized personal biases to prevent distorting the data through subjectivity (Miller, 2011; Tirgari, 2012; Yin, 2009).

Qualitative studies must demonstrate validity and reliability even though there is not a universal approach to validate qualitative studies (Creswell, 2003; Tirgari, 2012; Yin, 2009). Validity was accomplished by undertaking the study through a rigorous process to produce credible data (Cilesiz, 2011; Yin, 2009). The researcher achieved accuracy, dependability, and credibility by ethically ensuring the data collection procedures are stringent and capable of producing accurate information. Applying a rigorous process and methods for data collection and data analysis enhanced the transparency of the research, so readers were able to retrace the steps to reach similar findings (Cilesiz, 2011; Yin, 2009). Authenticity and credibility were mechanisms to strengthen the validity of the research, which the researcher achieved through member checking, conducting multiple interviews if necessary, and digitally video recording the interviews, and observation notes (Gordon et al., 2013; Zabloski & Milacci, 2012).

Validity and reliability were accomplished by

digitally video recording interviews with approval from
research participants, transcribing data, and allowing
the participants to member check the transcripts
(Tirgari, 2012). Triangulation was achieved by using
flight logbooks, observation notes, and interview data
were used to strengthen validity (Miller, 2013; Yin,
2009; Zivkovic, 2012). Data trustworthiness in
qualitative research occurred by combining
dependability, credibility, transferability, and confirmability (Baxter & Jack, 2008; Shank, 2006; Yin, 2009).
Credibility and dependability was achieved by using
the following procedures: (a) obtaining rich
descriptions, (b) identify the researcher's role and any
prejudices, (c) using purposeful and snowball
sampling strategies, (d) adhere to a strict qualitative
data collection and analysis method, (e) use member
checking, and comparing the cases after the data
analysis (Creswell, 2003; Gao, 2012; Miller, 2011).
Validity was achieved by reporting truthfully and
accurately (Shank, 2006). Member checking was
used to achieve integrity and credibility. Transparency
was essential this qualitative case study, so the
researcher listed the subjectivities, draft
comprehensive narratives of the study, develop an

effective and traceable data analysis plan, construct an audit trail, and included the limitations of the study that increased the transparency of the study (Cilesiz, 2011).

One reason for pursuing this qualitative single descriptive case study with three embedded units of analysis was to use human science inquiry to revisit experiences of general aviation pilots regarding the integration of TAA in general aviation (Creswell, 2003; Miller, 2011; Yin, 2009). Using qualitative research provides a robust platform to understand human actions that led to reductive reasoning and comprehension of the integration of TAA in general aviation (Miller, 2011; Yin, 2009). The explorative nature of qualitative research allowed the gathering of data from general aviation pilots accompanied by technical reports, and observation notes taken during the interview to determine the impact of integrating TAA in general aviation (Baxter & Jack, 2008; Ring, Ruth, & Ritchie; 2011; Yin, 2009). Quantitative research provided thorough numerical data; however, qualitative research provided a construct to investigate the meanings and descriptions of experiences by obtaining meanings, essences, and structures from first-hand occurrences that quantitative research lacks (Parylo et al., 2012; Zabloski & Milacci, 2012). The researcher used semi-structured interviews with opened-ended questions to collect data by interviewing research participants that were selected using purposeful sampling (Gordon et al., 2013; Levasseur, 2011). The aim of this research was to explore the integration of TAA in the general aviation sector through a qualitative single case study approach.

Population.

The purpose of this qualitative single descriptive case study was to explore the experiences of general aviation pilots on the integration of TAA and to determine how the automation of TAA can be used to increase aviation safety and improve pilot decision-making skills. General aviation pilots in Florida, Georgia, and South Carolina were interviewed. The research population was from the 77,842 general aviation pilots in Georgia, Florida, and South Carolina ("("FAA Certificated Pilots", 2011). In order to conduct face-to-face interviews, the targeted population was limited to general aviation pilots in Georgia, Florida, and South Carolina with glass cockpit experience. Based on the research design and method, the participants were placed into one of the following three embedded units of analysis or participant groups (a) general aviation pilots, (b) instrument and commercial rated pilots, and (c) certified flight instructors. The established criteria were for the general aviation pilots to have experience with TAA and recent involvement within the general aviation flying operations. The screening of flight logbooks was necessary to verify the pilots' flight qualifications and for appropriate placement into an embedded unit of analysis or participant group.

To participate in the study, the general aviation pilots had to have recent affiliation with general aviation; have a private pilot certificate at a minimum, at least 10 flight hours in TAA, and the authorization to fly as the sole occupant of the aircraft. Having 10 flight hours ensured the pilots have the minimum knowledge to manipulate the aircraft. Therefore, it was necessary to review the pilots' flight qualifications and logbooks. The goal was for the research participants to have different experience levels to

provide explanations on individual flying experiences
with glass cockpits in order to gain an in-depth
understanding on the impact of general aviation pilots
and to determine the applicable use of automation to
improve aviation safety and pilots' decision-making
skills. Therefore, meticulous adherence to the
selection criteria was essential to gain invaluable
insights on the phenomenon.

Sample.

The purposeful sampling method guided efforts
to target general aviation pilots to ensure the
participants had glass cockpit experiences in order to
contribute to understanding the phenomenon
(Gordon, Nichter, & Henriksen, 2013). Additionally,
purposeful sampling ensured the research
participants were qualified to provide in-depth
narratives to understand the integration of TAA in
general aviation. The researcher targeted participants
by sending official notifications to general aviation
organizations, flight schools, pilot organizations,
colleges and universities, and flight business offices in
Georgia, South Carolina, and Florida.

The advantage of using purposeful sampling was to ensure all research participants had credible experiences with the phenomenon and can provide reflections on the experiences (Suri, 2011). Qualitative interviews were typically 60 minutes in length and involved 11 to 13 research participants for each embedded unit of analysis or participant group (Levasseur, 2009; Yin, 2009). The three embedded units of analysis or participant groups were (a) private pilots, (b) commercial and instrument rated pilots, and (c) certified flight instructors; therefore, the researcher targeted 10 research participants for each embedded unit of analysis. The targeted sample was 35 research participants, 11 to 13 pilots for each embedded unit of analysis from a population of 77,842 general aviation pilots in Florida, Georgia, and South Carolina). In qualitative research, the number of research participants varies, so for this study there were 35 research participants. The interviewing continued until reaching a saturation point. Data saturation occurred when data collection did not reveal any new themes, perspectives, insights, or information (Suri, 2011). Because the study focused on the impact of TAA in general aviation, the population was limited to general aviation pilots with glass cockpit experience. Data saturation was achieved with a sample size of 35 research participants (Creswell, 2009; Goffin et al., 2012; Marshall et al., 2013); the targeted goal for this study was 30 research participants. For this study, it was imperative to ensure each embedded unit of analysis reached data saturation. The interviewer continued interviewing until all emerging themes were identified. Ultimately, data saturation determined the sample size by revealing all plausible themes (Goffin et al., 2012; Marshall et al., 2013).

Purposeful and snowball sampling were used to target general aviation pilots in Georgia, South Carolina, or Florida. Purposeful sampling was useful in determining research participants that were best suited to provide comprehensive insights on the integration of TAA in general aviation (Cane et al., 2010; Creswell, 2009; Suri, 2011). Purposeful sampling guided efforts to target a sample population to include general aviation pilots to ensure the participants have experiences in order to contribute to understanding the phenomenon (Suri, 2011). The purposeful sample were 35 research participants from a population of 77,842 general aviation pilots in

Florida, Georgia, and South Carolina ("FAA Certificated Pilots", 2011).

Snowball sampling in conjunction with purposeful sampling ensured the most credible and appropriate participants were targeted (Baltar & Brunet, 2012). Snowball sampling was a method in which one research participant provides the name and credentials of another potential research participant that meets the sample criteria, which happens continuously (Baltar & Brunet, 2012). To overcome geographical and time barriers associated with snowball sampling (Baltar & Brunet, 2012) this study was limited to the 77,842 general aviation pilots in Georgia, Florida, and South Carolina ("FAA Certificated Pilots", 2011). Snowball sampling provided the basis for targeting general aviation pilots with TAA experience to participate in the study.

Materials/Instruments.

The purpose of this qualitative single descriptive case study was to explore the experiences of general aviation pilots on the integration of TAA and to determine how the automation of TAA can be used to increase aviation safety and improve pilot decision-making skills. The researcher distributed the Research Study and Participant Notification (Appendix D) to recruit research participants at colleges and universities with general aviation flight programs, flight schools, flight business offices, general aviation pilots, and flying clubs in Georgia, South Carolina, and Florida. Interested research participants emailed or called the researcher to inquire about the study. Fifty-seven general aviation pilots in Georgia, South Carolina, and Florida responded to the Research Study and Participant Notifications. The researcher responded to each potential research participant via an introductory email or phone call to determine if the research participant met the minimum requirement of 10 hours of flight experience in TAA as the sole manipulator. The research participants that met the minimum criteria received a follow-up phone call from the researcher to schedule a face-to-face interview. The researcher explained to the potential research participants the rationale of the study, the informed consent process, and the minimal risk associated with the study.

The researcher used an Interview Question Guide (Appendix B) to ask open-ended questions during face-to-face interviews to collect data on the experiences of general aviation pilots regarding the integration TAA and how to use automation to improve pilots' decision-making skills. Face-to-face interviews allowed the researcher and interviewees to develop a rapport before the interviews started. Face-to-face interviews were the most appropriate interview technique, which allowed observation of the research participants' body language, facial expressions, and non-linguistic behavior during the interview (Cachia & Millward, 2011). The researcher did not conduct telephonic interviews because of the tendency for the interview to sound conversational and the inability to capture observations (Cachia & Millward, 2011; Kvale & Brinkman, 2009). The researcher asked open-ended questions from an Interview Questions Guide in a semi-structured format, which was supported by allowing the restructuring of previously asked questions to ensure the research participant was afforded the best opportunity to answer the question (Cachia & Millward, 2011). Asking open-ended questions afforded the research participant the opportunity to articulate experiences; thus, provided more reliable details and thick descriptions (Giles, Smythe, & Spence, 2012; Talbert, 2012).

To understanc the impact of integrating TAA in general aviation, the researcher interviewed general aviation pilots' with first-hand knowledge of operating TAA. To participate in this study, the research participants needed a private pilot certificate at a minimum, and at least 10 flight hours in TAA. Having 10 flight hours ensured the pilots have the minimum knowledge to manipulate the aircraft. Therefore, the interviewer verified the pilot's qualifications by reviewing their flight logbook. The data were collected from general aviation pilots with TAA experience from the states of Georgia, South Carolina, and Florida.

The research participants were asked the same questions as indicated in the Interview Question Guide (Appendix B). Cross-referencing notes taken during the interviews with interview data, observation data taken during the interview, and member checking strengthened the truthfulness and credibility of the data. Another method to strengthen reliability in qualitative research was to focus extensively on coding, which was a systemized process of grouping and breaking data into themes for interpretation during the data analysis phase (Mangioni & McKerchar, 2013). The datum were put through several iterations of categorizing, coding and triangulating for accuracy to ensure the outcome was reliable, which strengthened the validity of the data (Mangioni & McKerchar, 2013; Yin 2009).

Instrument Pilot Test.

A pilot study was conducted to evaluate the credibility and reliability of the Interview Question Guide. The general aviation pilots that participated in the pilot study were: a private pilot, a commercial and instrument rated pilot, and a certified flight instructor, which included a representative of each participant group. The objectives of the pilot study were: (a) to rehearse the interview protocols, (b) to develop interviewing skills, (c) to test the instrument (Interview Question Guide), (d) to refine the interview process, (e) to align data collection procedures, and (f) to implement processes to improve the outcome of the study (Kim, 2010). The pilot study interviews allowed a private pilot, a commercial and instrument rated pilot, and a certified flight instructor to assess the effectiveness of the Interview Question Guide, which indicated that the interview questions and the interview process could generate enriched data to determine the impact of integrating TAA in general aviation and how to leverage the advanced automated capabilities to improve pilots' decision-making.

Figure 19 - Primary Flight Display Garmin G1000; photo courtesy of Caricato da Nubifer.

Interviewing the pilot study participants provided recommendations that led to minor changes to the Interview Question Guide (Appendix B). One change reduced the number of questions on the Interview Question Guide from 17 to 15 questions to eliminate redundancy. Another modification resulted in modifying the pace of the interview, which provided more time for the research participants to respond to the questions. The final change was providing a copy of the Interview Question Guide to the research participant at the start of the interview, which allowed the study participants to read along with the interviewer, as well as reduced multiple readings of the question by the interviewer. The pilot study interviews provided the researcher with interviewing experience, the practice of interacting with the study participants, and the opportunity to gain experience in managing the interview. Additionally, the pilot study afforded the interviewer the ability to practice using follow-up questions and providing clarity on the research questions.

Data Collection, Processing, and Analysis.

The primary data collection sources were semi-structured interviews, technical reports and studies, flight logbooks, and observation notes. The interviews with general aviation pilots provided insight on the impact of integrating TAA in general aviation while the other collection sources conveyed historical, technical, and theoretical data that were corroborated with the interview data to holistically highlight the phenomenon (Berger et al., 2009). General aviation pilots were targeted as research participants through snowballing and purposeful sampling. Berger et al (2009) recommended using a four-phase approach for recruiting research participants. The phases involved determining initial contacts, screening, consenting, and enrollment and retention (Berger et al., 2009).

Determining Initial Contacts and Screening.

Determining initial contacts for potential general aviation pilots was achieved by providing a description of the Research Study and the Participant

Notification inquiry to flight schools, flight clubs,
colleges and universities with flight programs, to flight
business offices, and pilots with glass cockpit
experience. The Research Study and Participant
Notification were sent via mailed and emailed. The
Research Study and Participant Notification provided
details on the study and email and telephone contact
information for interested pilots. The notifications
directed interested general aviation pilots to contact
the researcher to determine if the pilot meets the
selection criteria for the study. To overcome the
logistical challenge of being in a different state,
general aviation pilots were allowed to call and
discuss their interests in the research and experience
with TAA to determine suitability to participate in the
study. Purposeful and snowballing assisted with
notifying general aviation pilots with adequate TAA
experience to participate in the study (Berger, Begun,
& Otto-Salaj, 2009).

Consent.

The consenting phase involved informing
interested general aviation pilots that met the criteria

to participate in the study of the informed consent

practices. The informed consent included the

following: (a) the purpose of the research, (b) the right

to withdraw from the study, (c) the procedures for

withdrawing, (d) that there will be no incentives for

participating in this study, (e) confidentiality

parameters, and (f) participant's rights (Berger et al.,

2009). The purpose of the informed consent was to

assist potential research participants with making an

educated decision on whether or not to participate in

the study. During the consenting phase, the potential

research participants reviewed and discussed the

informed consent form. Understanding the study and

the processes involved was essential before obtaining

written consent (Berger et al., 2009). Written consent

(Appendix C) was obtained before conducting any

interviews. The signed written consent forms were

filed and secured with the remainder of the research

documents that were safeguarded in a locked safe.

Enrollment.

After receiving approval from institutional

review board, the goal was to the recruit general

aviation pilots with TAA experience through purposeful and snowball sampling. Study notifications were disseminated to flying activities, flight clubs, colleges and universities with flight programs, flight business offices, and pilot organizations and associations to recruit qualified research participants. There are 77,842 general aviation pilots in Georgia, Florida, and South Carolina. During this process, one of the key goals was to develop a relationship with the general aviation pilots as well as provide information about the study, the consent process, the interview process, and measures to protect the research participants' privacy. Enrollment involved a relationship with the research participant and providing guidance to the general aviation pilots as needed throughout the study.

Data collection included face-to-face interviews with general aviation pilots, technical reports, and observation notes that were taken during the semi-structured interviews (Cachia & Millward, 2011; Yin, 2009). The collected data from the interviews was triangulated with technical reports, flight logbooks, and notes taken during the interviews (Baxter & Jack, 2008; Yin, 2009), because the accuracy and trustworthiness were increased via multiple sources (Patton, 2002; Zivkovic, 2012). Specifically, the technical reports and studies included: (a) the Aircraft Owners and Pilot Association's Air Safety Foundation (ASF) TAA Safety and Training Reports (2007), (b) the ASF Nall Report 2013, the NTSB Report on the Introduction of Glass Cockpits (2010), (c) the Government Accountability Office (GAO) Report on General Aviation Safety (2012), (d) the GAO Report on Initial Pilot Training (2011), and (e) a 2013 study on reducing general aviation accidents by utilizing airline operational strategies. The ASF Reports (2005, 2007) provided: (a) a historical perspective on the integration of TAA in general aviation, (b) over-reliance concerns, and (c) statistical data on aviation accidents involving TAA. Additionally, the ASF Reports provided details on: (a) safety implications, (b) TAA accident history, (c) required training for TAA, (d) human factors, and (e) software and hardware requirements for TAA. The ASF Reports provided invaluable that data answered the research questions in regards to training, automation, and pilot decision-making skills, as well as provided data for major theme development.

The GAO Reports on the introduction of glass cockpits and general aviation safety provided detailed information on: (a) two TAA quantitative studies, (b) the statistics breakdown of aviation accidents, (c) statistical data on the pilots involved in the accidents, (d) glass cockpit flight activity, (e) the status of general aviation safety along with areas for improvements, and (f) indications that TAA have not improved aviation safety. This report provided data on the research questions that inquired about flight training, aviation safety, and capitalizing on the technological capabilities of TAA. The 2013 study on reducing general aviation accidents by utilizing airline operational strategies provided a comparison between commercial and general aviation in order to eliminate accidents. The data within this report provided insight on strategies for improving aviation safety, flight training, safety management, decision-making, capitalizing on technology, pilot experiences, and accident prevention, which provided extensive data to respond to research questions and major theme development. The research participants' flight logbooks annotated demographic data on flight qualifications, flight experiences, and association with general aviation, which was essential when comparing the participant groups. The technical reports and TAA studies produced essential information during the data collection and data analysis phases that resulted in identifying major themes.

Open-ended questions were suitable for a qualitative case study methodology to allow participants to provide accounts of experiences with TAA (Tirgari, 2012). Asking open-ended questions in a semi-structured format was the best approach to capture the general aviation pilots' flight experiences and affiliation with general aviation (Allen, 2012). The interviews lasted between 47 to 69 minutes with the first 10 minutes used for introductions. The remainder of the interview session was based on questions from the interview question guide to keep the session focused and detailed in order to gain narratives on the research participants' experiences with TAA (Cilesiz, 2011). During the interview, the goal was to gather in-depth details on the research participant's first-hand knowledge on the integration of TAA in general aviation by asking to the participants to reconstruct specific events (Cilesiz, 2011). During the interview, each general aviation pilot was asked to provide meaning to individual experiences by evaluating the phenomenon from a present day lens (Cilesiz, 20116; Yin, 2009). The purpose of this was to provide a reference to how the phenomenon was progressing.

The following steps were the standardized procedures for conducting semi-structured interviews: (a) confirmed the appointment with the research participant 48-hours in advance via telephone followed by an e-mail with the location and directions to the interview site, (b) arrived at the airport one hour in advance to prepare for the interview, (c) pretested the primary and backup recording devices, (d) met and exchange introductory information with the research participant (e) allowed the research participant to read the informed consent form then explained the informed consent form, voluntary participation rights, the right to withdraw, and the purpose of the study (f) informed the participant that the interview was recorded and transcribed, (g) explained the pseudonym protocol, (h) answered any questions or concerns from the research participant before starting the interview, (i) placed the digital audio recording device on the lapel of the research participant and record the interview (j) concluded the interview, remove the digital video and audio recording equipment, and answered any questions or concerns of the research participant (k) transcribed the interview and allowed the participant to member check transcript within 5 days and return via e-mail, (l) sent a thank you note to the research participant for participating in the study, and (m) reviewed member checked transcripts.

Data processing involved transcribing the digitally video recorded semi-structured interviews, field notes, and observations into a word document (Brooks, 2010). Transcribing the interviews was a vital task (Brook, 2010). The digitally video recording device used for recording the interviews was used in conjunction with voice recognition software in order to transcribe the interviews expeditiously. Qualitative researchers recommended using voice recognition software with a digital video recorder to simplify transcribing the interviews into a document (Brooks, 2010). The transcripts underwent extensive scrutiny for accuracy and quality assurance by checking every sentence with the digitally video-recorded version of the interview.

The semi-structured interviews were triangulated with technical reports and studies, flight logbooks, and notes taken during the interview sessions (Gao, 2012; Yin, 2009). These multiple sources of data underwent rigorous scrutiny by the researcher to determine the major themes and patterns, which increased accuracy and credibility of the case study (Baxter & Jack, 2008; Yin, 2009; Zivkovic, 2012). Another purpose for triangulating was to identify data analogous in multiple sources, which assisted in discovering the major themes (Yin, 2009; Zivokovic, 2009).

Employing a modified van Kaam approach for data analysis in this qualitative single descriptive case study captured the impact of integrating TAA in general aviation (Cilesiz, 2011; Randles, 2012; Yin, 2009). The modified van Kaam approach was robust and proven, which enabled the capitalization on the modified structure process because this study had the potential to generate an extensive amount of data (Tirgari, 2012). Data analysis for this qualitative case adhered to the strict guidelines to include, "The exhaustion of resources, the emergence of regularities, and overextension" (Tirgari, 2012, p. 6).

The goal was to bracket the questions, remove all biases during the study, and conduct qualitative research to gather data on the experiences by using open-ended questions coupled with notes taken during the observation, pilots' flight logbooks, and technical reports and studies (Yin, 2009). A modified van Kaam approach was used to analyze the data (Creswell, 2003, Miller, 2009) (see Appendix A). The modified van Kaam model included the following steps: (a) digitally video recorded the interviews, (b) transcribed the interviews, (c) member checked the transcripts, (d) placed the transcripts in a logical flow, (e) coded the transcripts and all resources used for data collection, (f) repeated the coding process and dividing the data into phrases and categories, (g) horizontal themes, (h) developed textual and structural descriptions, (i) writing summary of each interview, (j) conducted second interview (if necessary), (k) conceptualizing and consolidating themes, (l) compared the units of analysis, (m) wrote a case summary, and (n) reviewed for an audit trail (Hycner, 1985; Tirgari, 2012). These steps formed the foundational framework that was used to conduct content analysis, which primarily involved identifying, coding, categorizing, classifying, and theming to convey the impact of integrating TAA in general aviation (Tirgari, 2012; Yin, 2009).

An essential step in data analysis was digitally video-recording all interviews (Tirgari, 2012) to enhance accuracy, reliability, and validity. Transcribing the interviews was an important step in preparing the interview data for processing and analyzing. Cilesiz (2011) recommended listening to the recorded interview several times to ensure the data is transcribed correctly. According to Cilesiz, one of the first steps in data analysis is to bracket any preconceived notions of the phenomenon. During the data reduction phase, the researcher read the transcripts, technical reports and studies, and flight logbooks multiple times and highlighted all statements in the transcript pertinent to the phenomenon (Creswell, 2003; Cilesiz, 2011; Yin, 2009).

Horizonalization was accomplished by coding, which was the process of identifying salient phrases, verbiage, and perspectives to classify and categorize into themes (Tirgari, 2012). The importance of establishing a classification and coding systems was pertinent to identifying, classifying, and labeling the patterns of the contents to develop meaning from the collected data (Yin, 2009; Zabloski & Milacci, 2012). The participants' flight and aviation expertise and experience varied, making it essential to establish a classification and coding system to identify and catalog the empirical data that highlighted the phenomenon, which enhanced credibility and confirmability (Patton, 2002; Trochim & Donnelly, 2008). The coding and classification methods kept efforts focused on the significant information (Levassuer, 2011; Patton, 2002).

Figure 20 – Glass Cockpit; photo courtesy of Larre-Anthony.

After highlighting the essential statements, the researcher transformed the data into statements and interpreted the meaning through the views of the research participants by breaking down statements into meaningful units (Cilesiz, 2011; Yin, 2009). Cilesiz postulated that horizonalization is the practice of giving every statement equal importance. Hyncer (1985) called this step a crystallization and condensation process because it requires the researcher to apply a general meaning to the words, phrases, and expression to include the same application to the invariant qualities and themes. The delimiting horizons or meanings process involved identifying statements inconsistent with participants' experiences (Creswell, 2003). All inconsistent statements made by the research participants were eliminated. Next, the researcher developed individual contextual descriptions by highlighting pertinent experiences, which reduced inconsequential statements, and then clustered and contextual salient themes, developed verbatim contextual from the interviews to support the construction of structural descriptions, and integrated singular patterns into clusters of similar (Tirgari, 2012; Yin, 2009).

Clustering the horizons in themes and organizing the themes assisted with developing individual contextual descriptions (Tirgari, 2012; Yin, 2009). Individual contextual descriptions were used to capture the research participants' experiences by categorizing and structuring the statements to form a narrative to describe the event (Cilesiz, 2011). The researcher developed individual contextual descriptions for each research participant.

The next step of the data analysis process was the development of structural descriptions. Structural descriptions were extensions of the individual contextual descriptions used to form structures based on the research participants' experiences (Cilesiz, 2011; Yin, 2009). Individual structural experiences captured the fundamental significance of how the research participants experienced the phenomenon (Cilesiz, 2011; Yin, 2009). The development of individual structural descriptions organized the narratives of the experience, themes, and essences that reflected the participant's own descriptions (Cilesiz, 2011; Yin, 2009). Individual structural descriptions represented the participants' personal details of the experiences, meanings, and descriptions (Cilesiz, 2011; Yin, 2009).

The next phase of the data analysis plan was synthesis, which pertained to classifying the analogous characteristics of the research participants' textures by grouping similar shared meaning units, in which the researcher assembled the shared meaning units into a composite contextual description (Cilesiz, 2011; Yin, 2009). The next step was identifying common structures from the individual structural description that developed into a composite structural description, which was a narrative that included frequent structures of the research participants' experiences (Cilesiz, 2011). After writing the composite description, the researcher produced an explanation of the phenomenon. The cases underwent exploration and comparison of the units of analysis to produce a case report, which was the final summary that provided comprehensive details on the phenomenon (Rao, 2013; Yin 2009).

Leedy and Ormrod (2005) postulated there was no exact structure for a case study report. The report included the research problem, research questions, an explanation of the experiences, the methods of collecting data and conducting analysis, the conclusion, developing a case report, and how the impacts affected technological determinism. The goal was for readers to have in-depth understanding of the impact after reading the case report (Yin, 2009).

The primary task during the data analysis phase was to discover clusters, themes, and patterns from the descriptions of the experiences (Tirgari, 2012; Yin, 2009). The context of the interview information underwent transcription, review for accuracy, and analyzed to capture the research participants' interpretation. The data was clustered into groups and sub-groups and compared to lead to final categories. According to Tirgari (2012), "the composite description summarizes the meaning of each major theme developed during the explication of the study data. The composite description expands on the data clustering and theming section" (p. 9). Using the modified van Kaam model supported the development of rich descriptions of the phenomenon based on the accounts of people with first-hand experiences (Tirgari, 2012).

Assumptions.

A major assumption was the general aviation pilots provided accurate accounts and details during the interviews. At the beginning of every interview, each general aviation pilot was instructed to answer the questions honestly and accurately. Each general aviation pilot was directed to ask for clarification on any question that was confusing in order to provide accurate descriptions and responses. Additionally, follow-up and probing questions were asked to ensure the best responses are captured. Another assumption was the general aviation pilots understood the questions and responded with appropriate answers and not hearsay. Allowing the pilots to member check the transcripts provided an opportunity to restate a previous response or ask for further clarification, which ensured the best responses were provided. The final assumption was this qualitative descriptive single case study provided rich details to develop an in-depth understanding of the impact of integrating TAA in general aviation and how to use the advanced capabilities to improve pilots' decision-making skills.

Limitations.

Limitations were factors the researcher does not control. A major limitation was this research study was limited to general aviation pilots in Georgia, South Carolina, and Florida, which was a small representation of the general aviation pilot population. Because of the different flight training programs and the small-targeted population, there was a possibility the research participants' experiences with TAA were not diverse enough to capture the holistic impact of integrating TAA in general aviation.

Another limitation was gaining the trust of the research participants so each could speak freely without fearing reprisal when answering the research questions. The research participants were briefed on the confidentiality and anonymity practices in order to attain factual and unfettered information from the research participants by not focusing on reprisal. Therefore, it was essential to establish a solid and open rapport with the research participants.

Given the scope and nature of the study, restricting the study to general aviation pilots was a limitation because corporate, military, and airline pilots have significant experience with flying TAA. This study focused primarily on general aviation pilots. By limiting the targeted population to general aviation pilots restricted the data and may have prevented gaining a holistic perspective on TAA. Research participants failing to meet for interview sessions was a limitation; thus, requiring the researcher to recruit another participates, which was time consuming and delayed data collection efforts.

Another limitation was the number of female participants. Only three female general aviation pilots participated in the study, which may have limited the outcome of the study.

A threat to the validity of the study was the primary researcher's personal experience with TAA in general aviation. Bracketing personal beliefs and experiences during the interviews was necessary to not to impede or affect capturing the experiences of the general aviation pilots. Bracketing occurred throughout the study to prevent personal biases or beliefs from having an impact on the study. The use of rich details and member checking were used to improve validity and credibility (Hathorn et al., 2009).

Delimitations.

Delimitations were boundaries established for the study the researcher controlled. One delimiting factor was restricting the study to general aviation pilots in Florida, Georgia, and South Carolina. The reason for limiting the study to general aviation pilots in Georgia, Florida, and South Carolina was to reduce the extensive travel and financial obligations in order to conduct face-to-face interviews. Another delimiting issue was restricting the interviews to 60 minutes to capture pertinent data. A flight qualification criterion was established and used to ensure the research participants had a private pilot certificate of a minimum of at least 10 flight hours in TAA, and authorization to fly as the sole occupant of the aircraft. Having 10 flight hours ensured the pilots had the minimum knowledge to provide credible and valid data.

Ethical Assurances

The application for human participation in the study required approval from Northcentral University's

IRB prior to collecting any data. The IRB reviewed all aspects of the regarding human participation in the study to ensure compliance with federal regulations as set forth by the Belmont Report (Smale, 2010). The Belmont Report provides ethical guidance primarily in three areas: (a) respect for human subjects, (b) beneficence, and (c) honesty (Smale, 2010). Each general aviation pilot was informed that harm or endangerment was not authorized during the study. There were three functional areas regarding respect to the humans that the researcher adhered to: (a) informed consent, (b) confidentiality of participants, and (c) autonomy of the participants (Rutherford-Hemming, Frances, & Rogers, 2012).

According to Plankey-Videla (2012), "The procedure of informed consent is required by professional codes of ethics and the International Review Board, draws from the principles of autonomy, beneficence, and justice" (p. 3). Each research participant signed an informed consent form (see Appendix C) before conducting any form of data collection. The purpose of the informed consent form was to provide the research participants with an explanation of the data collected and that ethical

assurance prohibits sharing data to sources external
to Northcentral University.

Confidentiality was a priority when conducting
research. The purpose of confidentiality was to
ensure the research participants were protected
throughout the study (Bozeman, Slade, & Hirsch,
2009). The research participants' identities and
personal information remain confidential. To ensure
confidentiality was maintained, all notes taken during
the interviews were stored in a safe that was locked.
The researcher briefed all participants on the
procedures to safeguard one's personal identity,
which was accomplished by giving each research
participant a pseudonym to safeguard identities.
Autonomy was based on the participant having a right
to withdraw from the research. Research participants
were advised of the right to withdraw from the
research at any time.

This qualitative case study required the
research participants to provide interpretations of
personal experiences with the phenomenon. From an
ethical perspective, the questions were in accordance
with the prescribed interview question guide and a
professional rapport with the participant that was

maintained throughout the interview. Compliance was accomplished by maintain a professional demeanor and working within the rules set forth by the IRB. Bracketing, epoché, and accurately annotating the data were methods used to remain in compliance with the IRB guidelines.

Summary

The purpose of this qualitative single descriptive case study was to explore the experiences of general aviation pilots on the integration of TAA and to determine how the automation of TAA can be used to increase aviation safety and improve pilot decision-making skills. The results of this study provided an in-depth understanding of the phenomenon, which could lead to theoretical changes in implementing technologies in general aviation. A qualitative case study allowed the gathering of information from general aviation pilots through interviews and document analysis to determine the themes to understand the phenomenon.

Undertaking this study through a quantitative design process was not feasible because a quantitative approach does not support exploring human experiences by gathering data on research participants' experiences (Cliesez, 2011). The measurement and assessments provided analytic data to develop a comprehensive understanding on integrating TAA in general aviation and highlighted the inconsistencies with current theoretical beliefs on implementing technologies in general aviation.

CHAPTER IV
FINDINGS

The purpose of this qualitative single descriptive case study was to explore the experiences of general aviation pilots on the integration of TAA and to determine how the automation of TAA can be used to increase aviation safety and improve pilots' decision-making skills. This descriptive single case study used three embedded units of analysis to develop an in-depth understanding of integrating. The theoretical foundational was based on the technical determinism theory, which revealed significant aftereffects in flight-training, aviation safety, aeronautical decision-making (ADM), and regulatory problems that emanated from the integration of TAA in general aviation (McCracken, 2011). The research questions included one central question and three sub-questions that developed the underpinning for collecting data from general aviation pilots, which were integrated with supplemented data from technical reports and studies, and observation notes. The questions from the Interview Question Guide aligned to the central research question and sub-questions. This descriptive case study was adequate to organize the case and present the results in an effective manner (Yin, 2009).

This chapter highlights the results of the pilot study, results of the study, the evaluation of findings, the findings in the context of the theoretical framework, and the summary. This chapter includes of three major sections; the first segment addresses the study results to include the pilot study results and the findings of the research questions. The section addresses the data analysis results such as: (a) the data collection methods to identify salient themes, (b) the triangulation of semi-structured interviews, technical reports and studies, and observation notes, (c) an embedded units of analysis comparison, and context for the theoretical framework. The second section provided the details of the findings, which will address the findings through the technological determinism lens. The final section summarized the critical points of the chapter.

Results.

The case study results underwent a rigorous process to discover the salient themes by scrutinizing the collected data for each research question and the supplemented technical reports and studies. The purpose of the data collection and data analysis phases were to explore the personal experiences of general aviation pilots and the analysis of the technical reports and studies, and observation notes to attain an in-depth comprehension on the integration of TAA in general aviation and how to leverage the advanced capabilities of glass cockpits to improve aviation safety and pilots' decision-making skills. There were a total of 38 interviews to include three during the pilot study with private pilots, commercial and instrumented rated pilots, and certified flight instructors. Eleven private pilots, 13 commercial and instrument rated pilots, and 11 certified flight instructors were interviewed for this case study.

Results of the instrument pilot study. To ensure the Interview Question Guide and interview process were suitable, the researcher conducted an instrument pilot study with three study participants. The three pilot study participants included a private pilot, a commercial and instrument rated pilot, and a certified flight instructor, which included a representative for each embedded unit of analysis. The objectives of the pilot study were: (a) to rehearse the interview protocols, (b) to develop interview skills, (c) to test the instrument (Interview Question Guide), (d) to refine the interview process, (e) to align data collection procedures, and (f) to implement processes to improve the outcome of the study (Kim, 2010).

Research participants for the pilot study were recruited by employing purposeful sampling. The researcher used purposeful sampling to recruit participants for the pilot study to ensure the participants were general aviation pilots with TAA experience to provide credible and reliable data to help understand the impact of glass cockpits in general aviation. The researcher mailed, faxed, and e-mailed the Research Study and Participant Notification Letters (Appendix D) to flight schools, flight business offices, colleges and universities with flight programs, an aviation safety research organization, general aviation aircraft manufacturing companies, a collegiate aviation organization, several pilot organizations, a flying club in northeast Georgia, a flight school in central Florida, and general aviation pilots to recruit participants for the pilot study. During the initial phone exchange, the researcher explained the procedures listed on the Informed Consent Form (Appendix C) and arranged the interview date, as well as emailed a copy of the Informed Consent Form to the research participant before meeting with the interviewee, which were signed before commencing the interviews.

From the dissemination of the Research Study and Participant Notification Letters (Appendix D), the researcher received 23 phones or emails from general aviation pilots that were interested in participating in the pilot study. The researcher followed up with the participants and selected a private pilot, commercial-instrument rated pilot, and certified flight instructor that had TAA experience as a general aviation pilot. The researcher used no specific selection criteria for selecting the three pilots other than having TAA experience as a general aviation pilot with the required pilot certification. The participants that were not selected for the pilot study were sent an e-mailed by the researcher to inform them that quotas for the pilot study had been attained; however, their participation in the primary research study was needed. Eighteen of the 20 participants that volunteered for the pilot study participated in the research study.

Findings. Profiles of the pilot study participants. The researcher conducted three face-to-face interviews during a 6-day period. The private pilot (PP-1) was a 21-year-old male from South Carolina, has 203 flight hours, 79 flight hours in TAA, and has been licensed as a general aviation pilot for three years. The commercial and instrument rated pilot (CIR-1) was a 43 year old male from Florida, has a 784 total flight hours, which 310 of the flight hours were in a TAA, and has been licensed as a general aviation pilot for 11 years. The certified flight instructor was a 32-year-old male from Georgia, has 4,217 total flight hours, 1,831 flight hours in TAA, and has been a licensed pilot for 12 years.

Table 1 - Pilot Study Participants Data

Participant ID	Interview Date	Total Flight Hours	TAA Flight Hours	State of Residence	Informed Consent Provided	Interview Length	Sex	Age
CFI-1	09/10/14	4,217	1,831	GA`	Yes	53:28	M	32
PP-1	09/12/14	203	79	SC	Yes	66:43	M	21
CIR-1	09/14/14	784	310	FL	Yes	67:47	M	43

n=3

The pilot study participants met the criteria of having a private pilot license, at least 10 hours in TAA, and the authorization to operate the aircraft as the sole pilot. Purposeful sampling was used to recruit three pilot study participants, which represented the three embedded units of analysis. This study had three embedded units of analysis: (a) private pilots, (b) commercial-instrument rated pilots, and (c) certified flight instructors. The researcher had to ensure each research participant met the criteria in order to attain credible data from the general aviation pilots.

Critical Findings.

Changes implemented from the pilot study.
The pilot study yielded several critical findings to
improve the quality of the instrument (Interview
Question Guide, Appendix B) as well as provided the
researcher with actual interviewing experience. The
responses from the pilot study participants resulted in
reducing the Interview Question Guide from 17 to 15
questions to eliminate redundancy in similarly worded
questions. The pilot study revealed the interviews
exceeded the allotted 60-minute period; hence, forced
the researcher to focus on the interview pace.
Reducing the number of questions from 17 to 15
aided in controlling the length of the interviews. The
following propositions were used to formulate the
research questions for the case study (Baxter & Jack,
2008; Yin, 2009):

(a) General aviation pilots value the integration of
     TAA in general aviation
(b) General aviation pilots require additional
     training to fly TAA proficiently and safely

(c) More governmental oversight and coordination
is needed for the integration of TAA, and

(d) General aviation continues to be plagued by
too many aviation accidents.

The proposition shaped the questions to
ensure the participants provided enriched data to
explore the integration of TAA in general aviation. The
final change was providing a copy of the Interview
Question Guide to the research participant at the start
of the interview, which allowed study participants to
read along with the interviewer, as well as reduced
multiple readings of the question by the interviewer.
The pilot study highlighted that adhering to the
interview protocol was critical to collecting essential
data from general aviation pilots. The pilot study
allowed the research participants to assess the
effectiveness of the Interview Question Guide, which
indicated the interview questions and the interview
process could generate enriched data to determine
the impact of integrating TAA in general aviation and
how to leverage the advanced automated capabilities
of glass cockpits to improve pilots' decision-making.

Private Pilots Demographic Data.

Eleven private pilots were interviewed for this particular group. These eleven pilots represented the private pilot embedded unit of analysis. The group was private pilots from Georgia, South Carolina, and Florida. There were nine male participants and two female participants. The age range for the participants in this group was from 21 to 61 years with an average age of 31.6 years. The group average of total flight hour experience was 729.3 flight hours, which was significantly higher because PP-3 had 3,309 and PP-6 had 1,343 flight hours; thus, increased the group's total flight hour average. Flight experience in TAA ranged from 18 to 2,194 flight hours with the average TAA experience of the group equated to 290.2 flight hours; however, the overall flight hours ranged from 18 to 3,309. The average TAA experience was higher because PP-3 had 2,194 TAA flight hours, which made the group's average TAA experience significantly higher. The group's average years of as a pilot were 5.72 years. All the private pilot research participants met the minimum requirements of having a private pilot license, a minimum of 10 flights in a

TAA, and the authorization to operate the aircraft as

the pilot-in-command. Table 2 lists the summary of

demographics for the private pilot group.

Table 2 - Private Pilot's Demographic Data

Participant ID	Interview Date	Total Flight / TAA Hours	State of Residence	Informed Consent Provided	Interview Length	Sex	Age	Years as Pilot
PP-2	10/01/14	115 / 83	GA	Yes	57:49	M	23	2
PP-3	10/04/14	3,309 / 2,194	GA	Yes	52:03	M	61	17
PP-4	10/03/14	257 / 210	FL	Yes	55:31	F	29	4
PP-5	10/03/14	682 / 107	FL	Yes	66:17	M	33	8
PP-6	10/05/14	1,343 / 58	SC	Yes	56:51	M	24	2
PP-7	10/06/14	231 / 29	SC	Yes	59:18	M	21	3
PP-8	10/12/14	871 / 32	SC	Yes	69:47	M	39	8
PP-9	10/19/14	746 / 286	GA	Yes	54:03	M	27	10
PP-10	10/21/14	174 / 85	FL	Yes	53:21	F	23	3
PP-11	10/21/14	91 / 91	FL	Yes	58:54	M	21	1
PP-12	10/23/14	203 / 18	GA	Yes	62:49	M	47	5

n=11

Figure 21 – Professional Pilot in the USA; Photo courtesy of Ali Rezaamidi.

Thirteen commercial and instrument-rated pilots participated in the study. The 13 participants provided interview data to support the commercial-instrument rated pilots embedded unit of analysis. The group was commercial and instrument rated pilots from Georgia, South Carolina, and Florida. The group included 13 male participants who all meet the minimum requirements of having commercial and instrument ratings, a minimum of 10 flights hours in TAA, and the authorization to operate the aircraft as pilot-in-command. The average age for the participants was 37.6 years old. The group's average years of flight experience were 12.5 years. The average TAA flight experience for the group was 293 TAA flight hours, which was considerably higher because CIR-5 had 1,583 TAA flight hours. The average flight hour experience for the group was 1290 flight hours. The average flight experience was substantially higher because CIR-5 had 4,667 total flight hours, CIR-6 had 2,265 total flight hours, CIR-7 had 1,428 total flight hours, CIR-11 had 1,013 total flight hours, and CIR-14 had a 2,754 total flight hours. Table 3 lists the summary of demographics for the commercial-instrument rated pilot group.

Table 3 - Commercial- and Instrument-Rated Pilots Demographic
Data

Participant ID	Interview Date	Total Flight / TAA Hours	State of Residence	Informed Consent Provided	Interview Length	Sex	Age	Years as a Pilot
CIR-2	10/08/14	328 / 47	FL	Yes	49:43	M	27	6
CIR-3	10/08/14	443 / 29	FL	Yes	63:03	M	47	11
CIR-4	10/12/14	951 / 214	SC	Yes	61:23	M	51	16
CIR-5	10/12/14	4,667 / 1,583	SC	Yes	53:29	M	57	22
CIR-6	10/16/14	2,265 / 219	GA	Yes	51:28	M	43	15
CIR-7	10/16/14	1,428 / 354	GA	Yes	63:37	M	26	8
CIR-8	10/19/14	487 / 73	GA	Yes	55:09	M	22	4
CIR-9	10/24/14	989 / 182	FL	Yes	67:58	M	32	13
CIR-10	10/27/14	259 / 200	FL	Yes	58:51	M	20	2
CIR-11	10/29/14	1,013 / 107	SC	Yes	52:43	M	38	15
CIR-12	10/29/14	663 / 375	SC	Yes	59:21	M	32	9
CIR-13	10/31/14	523 / 128	GA	Yes	61:38	M	41	14
CIR-14	10/31/14	2,754 / 303	GA	Yes	57:26	M	53	27

n=13

Eleven certified flight instructors participated in the study. The eleven participants provided interview data to support the certified flight instructors embedded unit of analysis. The group included certified flight instructors from Georgia, South Carolina, and Florida. There were 10 males and one female from this group that participated in the study who all meet the minimum requirements of having certified flight instructor rating, minimum of 10 flights hours in TAA, and the authorization to operate the aircraft as the pilot-in-command. The average age for the participants was 40.7 years. The average years of flight experience were 14.7 years and the average TAA experience for the group was 489 hours. The average flight hour experience was significantly higher because CFI-6 had 1,401 total TAA flight hours, and CFI-8 had 1,933 total TAA flight hours.

Figure 22 - Operators in a control room pilot and monitor video feeds from a remotely piloted UAV. Photo courtesy of Gerald Nino, CBP, U.S. Dept. of Homeland Security – CBP.

The TAA flight experience ranged from 22 to 1,933. The overall flight hour experience for the group was 2300.2 flight hours, which was significantly higher because CFI-3 had 1,711 total flight hours, CFI-4 had 3,832 total flight hours, CFI-6 had 6,429 total flight hours, CFI-8 had 5,462 total flight hours, CFI-10 had

2,397 total flight hours, and CFI-12 had 1,946 total

flight hours. Table 4 lists the summary of

demographics for the certified flight instructor group.

Table 4 - Certified Flight Instructors Demographic Data

Participant ID	Interview Date	Total Flight / TAA Hours	State of Residence	Informed Consent Provided	Interview Length	Sex	Age	Years as Pilot
CFI-2	10/04/14	872 / 136	FL	Yes	62:53	M	33	11
CFI-3	10/04/14	1.711 / 493	FL	Yes	56:56	M	41	14
CFI-4	10/08/14	3,832 / 296	FL	Yes	55:24	M	61	27
CFI-5	10/14/14	807 / 221	SC	Yes	57:11	M	37	8
CFI-6	10/14/14	6,429 / 1,401	SC	Yes	53:49	M	59	23
CFI-7	10/14/14	568 / 42	SC	Yes	56:57	F	29	7
CFI-8	10/17/14	5,462 / 1,933	GA	Yes	59:28	M	44	21
CFI-9	10/17/14	512 / 22	GA	Yes	58:14	M	26	4
CFI-10	10/17/14	2,397 / 471	GA	Yes	64:38	M	37	18
CFI-11	10/26/14	766 / 93	FL	Yes	57:29	M	38	15
CFI-12	10/26/14	1,946 / 273	FL	Yes	64:37	M	43	14

n=11

Results.

Thirty-five general aviation pilots were

interviewed over a five-week period from October

2014 to November 2014. The 38 general aviation

pilots included 11 private pilots, 13 commercial-

instrument rated pilots, 11 certified flight instructors,

and three general aviation pilots for the pilot study. Of

the 35 general aviation pilots, three were female. The small number of female participants limited the outcome of the study; thus, requiring additional research on the impact of TAA on gender. The data provided by the 35 research participants achieved data saturation as indicated by the lack of introduction of new information during the interview sessions. Each participant answered all the questions on the Interview Question Guide (Appendix B). The researcher provided results of the questions by embedded units of analysis, which were a private pilot group, a commercial-instrument rated pilot group, and a certified flight instructor group. The interview duration ranged between 49-69 minutes. Each semi-structured interviewed was digitally recorded with a Sony Video Recorder and each participant wore a lapel microphone on the collar to enhance the sound during the recording. The interviews were transcribed into a word document within 12-36 hours after the interviews. Each participant was afforded an opportunity to member check their transcripts which increased reliability, enhanced accuracy, and ensured the elicit data correctly articulated the research participants' experiences (Harper & Cole, 2012).

A modified van Kaam method was used to identify patterns, categories, and themes from the collected data and the content analysis from technical reports and studies (supplemented data). The collected data were processed in accordance with Appendix A to identify the major themes of each embedded unit of analysis to determine the impact of integrating TAA in general aviation. After identifying the major themes for each embedded unit of analysis, the researcher compared the results of each embedded units of analysis.

To determine the impact of integrating TAA in general aviation and to determine how the automation of TAA can be used to increase aviation safety and improve pilots' decision-making skills, the researcher developed one central research question and three sub-questions, and complemented the questions with the Interview Question Guide (Appendix B). The intended purpose of these questions was to form the foundation for analyzing data from the semi-structured interviews from private pilots, commercial-instrument rated pilots, and certified flight instructors and content analysis from technical studies and reports. The research questions guided the results of the study. The questions were designed to gather data on TAA, pilot decision-making, and automation in general aviation, and capitalizing on glass cockpits to improve aviation safety. Questions 1-5 of the Interview Question Guide were used to gather demographic and flight data on the participants. Below are the one central question, the three sub-questions, and the associated questions from the Interview Question Guide.

Q1. What are the experiences of general aviation pilots regarding the integration of TAA?

- How would you describe your flight training to prepare you to fly TAA?
- What changes can the government and industry make to better prepare general aviation pilots to transition between conventional aircraft and TAA?
- How do you view the regulatory approach in general aviation regarding TAA?
- How do you use aeronautical decision-making training in your general aviation flying?

Q2. How can general aviation pilots use the automation of TAA to reduce accidents?

- What changes can the government and industry make to better prepare general aviation pilots to transition between conventional aircraft and TAA?
- How do you use aeronautical decision-making training in your general aviation flying?
- Describe your experiences of how TAA has impacted your approach to flying.

Q3. How can general aviation pilots reduce the effects of technological innovations to capitalize on the positive attributes of TAA?

- What is the best approach to integrate technologies into general aviation?
- How do you feel about the safety implications caused by integrating TAA in general aviation?

Q4. What is the impact of TAA on aeronautical decision-making skills of general aviation pilots?

- What is your view on of FITS?
- How do you use Single Pilot Resource Management in general aviation?

Major themes of the study.

The seven prominent themes (see Table 6) identified during the data analysis were: (a) training, (b) safety, (c) proficiency, competency, and familiarization, (d) transformative and challenging transition, (e) situational awareness, (f) decision-making, SPRM, and ADM, and (g) regulation. The minor themes identified were: (a) limit innovations and (b) cognition. Table 6 is a complete listing and frequency of the themes from the interviews and document analysis after conducting thematic analysis. Table 5 is a breakdown and comparison of the themes by the embedded units of analysis. The major themes were determined by the most frequency (see Table 5). The top themes from each research question by the embedded units of analysis were compiled into Table 6 along with the major themes from the technical reports and studies. The researcher compiled all the themes from the data

analysis and selected the most frequent themes as

the salient themes. Themes were compiled based on

the embedded units of analysis. The researcher was

interested in the themes of each embedded units of

analysis to make comparison between the embedded

units of analysis. The research identified additional

themes and added the frequency of each theme to

the themes listed in Table 6 after conducting the

document analysis on the technical reports and

studies.

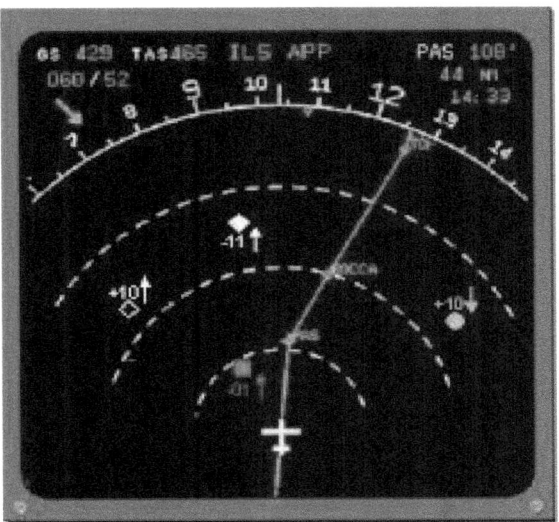

Figure 23 – Traffic Collision Avoidance System (TCAS) and
EHSI cockpit display; Photo courtesy of Jemr69.

Table 5 - Comparison of Themes by the Embedded Unit of
Analysis

	Private Pilot Group	Commercial Instrument Group	Certified Flight Instructor
RQ1	• TAA endorsement • No regulatory Change • Challenging Transition • Cognitively Demanding • Mandatory Training	• Regulatory Change • Comfort • Training • Limit Innovations	• Regulatory • Transformative
RQ2	• Situational Awareness • Discipline / Airmanship • No Regulatory Change • ADM / SPRM	• Situational Awareness • Training • Personal Standards	• Situational Awareness • Cognitive
RQ3	• Training • Proficiency • Familiarization • Limit Innovations	• Training • Safety • Pilot Centric	• Training • Limit Innovations
RQ4	• Situational Awareness • Decision-making • Training	• Safety • Competency	• SPRM • Cockpit Management

Themes from technical reports and studies.
The themes from the technical reports and studies
were included in Table 6. The technical reports and
studies supplemented the interview results. In a
technical report on the introduction of glass cockpits
in general aviation aircraft by the National
Transportation Safety Board (2010), the salient theme
was aviation safety. For example, the board members

declared that glass cockpits have not improved
general aviation safety as expected. The report
highlighted the FAA has not mandated FITS as an
approved training curriculum. The report also
indicated the aviation knowledge test does not
address glass cockpits or require the pilots to be
tested on glass cockpit instrumentation.

Table 6 - Compilation of Themes

Themes	Frequency of Themes
Training	9
Safety	8
Proficiency / Competency / Familiarization	6
Transformative / Challenging Transition	6
Situational Awareness	5
Decision-making / SPRM / ADM	5
Regulatory	5
Cognition	2
Limit Innovations	2

In a scholarly study conducted by McCracken
(2011), the major themes specified were transition
complexity and training. For instance, McCracken
indicated significant changes to the flight training
curriculums were necessary for pilots to learn new
scan patterns, interpret the data, and managing the
redistributed workload, which defies existing practices
and processes. The study revealed pilots struggled

and felt less comfortable with the transition from conventional aircraft to TAA even after completing two semesters of flight training. The study determined that pilots' overall progress declined due to the transition to TAA and the lack of instructor familiarity with the new aircraft (McCracken, 2011). This is indicative of the aftereffects of integrating TAA in general aviation. The participants in McCracken's study emphasized the lack of preparation and comfort in flying the TAA. For example, 70% of the participants felt TAA are more difficult to fly than conventional aircraft and that the participants focus extensively on the automation and neglected to maneuver the aircraft.

In a comparison study between two TAA by Franza and Fanjoy (2012) revealed themes on safety, training, and transitioning difficulty. For example, Franza and Fanjoy stressed pilots need to understand the advanced capabilities of TAA for associated risks and increased probability of being in an accident if not proficient in glass cockpits. The researchers recommended FITS as a method to reduce the risk associated with flying TAA by instructing pilots to leverage and integrate new technologies in their ADM to become safer pilots.

Garibay and Young (2013) conducted a study on reducing general aviation accidents by using airline strategies, which training, SPRM, and safety themes emerged from the data. Garibay and Young recommended for the general aviation community to incorporate cockpit technologies, offer incentives for pilots to undergo recurring safety training to eradicate risk, and to develop a check and balance system to increase the safety envelope. For the general aviation community to reduce the number of aviation accidents, the industry requires in-depth focus and training strategies to improve general aviation safety (Garibay and Young, 2013).

A major theme derived from the AOPA Nall Report (2013) was safety; the data specified there were 1,160 general aviation accidents and 334 fatalities in 2012 and 948 general aviation accidents and 283 fatalities in 2013. The data highlighted that personal flight operations was the lead category for the most accidents with 74% in 2012 and 76% in 2013 (AOPA, 2013). In 2013, 52% of the accidents occurred with an instrument rated pilot onboard while 23% of the accidents occurred with a certified flight instructor onboard. The number of accidents in 2012 and 2013 accompanied by the high percentage of aviation accidents, with instrumented rated pilots and certified flight instructors onboard, was indicative of the safety plight.

Private Pilot (PP) Group.

Research question one was, "What are the experiences of general aviation pilots regarding the integration of TAA?" Table 7 provides a summary of the responses to the question and the supporting questions from the Interview Research Guide.

Table 7 - Private Pilot Group Responses to Research Question 1 (RQ1)

Category of Response	Frequency	Percentage
TAA Endorsement	9	75.0
No Change to Regulation	8	66.7
Challenging Transition	8	66.7
Cognitive Demanding	7	63.6
Make FITS Mandatory / Advanced Training	6	54.5
Transformative	1	9.0
Unintuitive	1	9.0
Overwhelming	1	9.0

n=11

The participants answered the question to reflect their experiences with TAA. Majority of the participants (75%) responded the integration of TAA in general aviation should require some sort of endorsement indicating the pilot had completed specific training to fly TAA. For instance, PP-9 opined,

There was no existing requirement mandating a pilot to undergo TAA training or what was credible TAA training, this ambiguity has resulted in aviation accidents and fatalities. The endorsement would serve as verification that the pilot had received the necessary training to operate TAA.

Sixty-six percent of the private pilots responded that the transition from conventional aircraft to TAA was challenging. Similarly, 63% of the private pilots responded that flying TAA is cognitively demanding. For example, PP-2 noted, "The transition to TAA from conventional aircraft was exhaustive, demanding, and foreign." PP-10 stated, "The transitioning to TAA was unintuitive and required more mental focus and stamina." Interestingly, most of the private pilots (66%) declared that no changes to existing regulations were required due to the integration of TAA. PP-3 stated, "Existing regulations were overarching and thorough enough to include TAA." PP-9 avowed, "or enhance general aviation operations." Instead of implementing new regulations, most the general aviation pilots (54.5%) advocated for the enforcement of FITS and similar training programs. In fact, PP-5 stated, "FITS was the most advanced and thorough training I have completed. This type of training would save lives and keep pilots flying."

The above-mentioned results are substantiated by existing literature. In a previous study, researchers acknowledged the lack of a standardized training program for pilots transitioning from conventional aircraft to TAA (Lindo et al., 2012). Lindo et al (2012) reported the integration of advanced technologies in general aviation have negatively impacted aviation safety. Researchers indicated FITS is an effective training program that prepares general aviation pilots to fly TAA, even though FITS is not mandated as required training by regulatory officials (Lindo et al., 2012). The NTSB (2010) reported that TAA provided increased available safety; however, to achieve the increased level of safety requires pilots to undergo additional training such as FITS. Research indicated that aviation insurance companies require pilots to complete specific training requirements such as FITS or specific transition training in order to receive insurance coverage because the FAA leadership does not enforce FITS.

Research question two was, "How can general aviation pilots use the automation of TAA to reduce accidents?" Table 8 provides a summary of the responses to the question and the supporting questions from the Interview Research Guide. The participants answered the question to reflect their experiences with the automation of TAA. All of the participants (100%) responded that leveraging the automation of TAA enhanced situational awareness. For example, PP-1 noted, "Using the automation and the autopilot reduces fatigue and allowed to pilots to stay mental alert through all phases of flight." Similarly, PP-12 asserted, "The autopilot alleviate the stress of having to manually control the aircraft, which improved my cockpit management skills." Sixty-three percent of the private pilots responded that being discipline and practicing sound airmanship in TAA reduced aviation accidents. PP-8 stated, "Aircraft with glass cockpits fly themselves; the pilots just need to be discipline and operate the aircraft within the designed safety parameters and manage the automated data." Fifty-four percent responded the existing regulations mandated enough guidance to manage TAA in GA. For example, PP-2 asserted, "Changing the regulations would only add confusion and perplexity to the process. General aviation flying was meant to be fun and adding new regulations would hamper the recreational aspect of general aviation." Most of the private pilots (54.5 percent) responded that regular training in aeronautical decision-making (ADM) and single pilot resource management (SPRM) was imperative to reducing TAA accidents. PP-2 responded, "The general aviation community needed better training models, for those pilots operating complex aircraft that are single-piloted. Single Pilot Resource Management is general aviation version of Crew Resource Management but most general aviation pilots are not familiar with it." PP-3 stressed, "ADM saved my life on numerous occasions after experiencing and in-flight emergencies, and I had to rely on my training and started training regularly to stay proficient."

Table 8 - Private Pilot Group Responses to Research Question 2 (RQ2)

Category of Response	Frequency	Percentage
Situational Awareness	11	100.0
Discipline / Airmanship	7	63.6
No Changes to Regulation	6	54.5
ADM / SPRM	6	54.5
Flight Simulators	2	18.8

n=11

Research question three was "How can general aviation pilots reduce the effects of technological innovations to capitalize on the positive attributes of TAA?" Table 9 provides a summary of the responses to the question and the supporting questions from the Interview Research Guide. The participants answered the question to reflect their experiences with the automation of TAA. One hundred percent of the private pilot participants responded that training was the key to learning to leverage the capabilities of TAA to capitalize on the automation.

For example, PP-6 stated, "I had to invest in an additional 10 hours of flight training in TAA in order to fully understand how to maximize the use of the autopilot and navigational capabilities." PP-10 declared, "I had to repeat several flight lessons in the training program just to comprehend all the advanced automation." Sixty-six percent of the private pilots avowed that flying regularly improved proficiency and increased their confidence in operating TAA. PP-11 affirmed, "Flying every day while getting my private pilot license helped me develop the skills and technical understanding to fly glass cockpits." Similarly, PP-2 stated, "Flying along as the second pilot in my father's Cirrus every weekend was beneficial in learning to use all the automation and avionics."

A majority of private pilots recommended getting exposure with TAA or flying with someone else before flying as the sole pilot to become familiar with the aircraft. For example, PP-4 stated, "I flew over 30 hours in simulators before purchasing a Cessna 172 with a glass cockpit to get familiar with the equipment and flight handling characteristics of the aircraft." Some pilots (36%) recommended using simulators to increase familiarization. Fifty-four percent of the private pilots responded there are too many innovations integrated into general aviation. PP-12 stated, "I have plans to purchase a TAA but there were too many selections to choose from and I cannot

determine which automated technologies to include in

the cockpit."

Table 9 - Private Pilot Group Responses to Research Question 3 (RQ3)

Category of Response	Frequency	Percentage
Training	11	100.0
Stay Proficient	7	63.6
Familiarization	6	54.5
Reduce Innovations	6	54.5
Use Simulators	4	36.3

n=11

Research question four was "What is the

impact of TAA on aeronautical decision-making skills

of general aviation pilots?" Table 10 provides a

summary of the responses to the question and the

supporting questions from the Interview Research

Guide.

Table 10 - Private Pilot Group Responses to Research Question 4 (RQ4)

Category of Response	Frequency	Percentage
Improved Situational Awareness	10	90.0
Decision Making	9	81.8
Increase FITS Training	6	54.5
SPRM (new approach)	4	36.3

n=11

The participants answered the question to reflect their experiences with ADM. All of the participants responded that TAA increased the pilot's situational awareness and aeronautical decision-making abilities. PP-9 stated, "TAA enhanced my decision making skills because I can use the automation to fly the aircraft which is why I think through the critical phases of flight without sacrificing my situational awareness." PP-7 avowed, "Since transitioning to TAA my situational awareness and ADM have improved tenfold. I really don't worry about falling behind the aircraft, the glass cockpit was an enormous help in alleviating the mental math and processing when flying." Fifty-four percent of the private pilots reported that FITS were essential when undertaking their transitioning training. PP-6 stated, "FITS was the foundational training that taught me how to maximize my ADM skills when piloting TAA." Similarly, some of the private pilots (36.3%) stated that SPRM provided pilots with advanced skills to operate advanced aircraft as single pilots. PP-5 affirmed,

I learned of SPRM through a flying associate, my flight school does not teach it so I took flight lessons from a different school to gain exposure and it made me a better pilot with a methodical process to work through any situation when flying.

PP-12 avowed, "I learned SPRM from day one of my transition to TAA, it was a bit much at the beginning once I put all the pieces together it made me a better pilots."

Figure 24 - TCAS and IVSI cockpit display (monochrome); photo courtesy of Mattes.

Commercial-Instrument Rated (CIR) Pilot Group.

The pilots in this group have attained both an instrument and commercial certification. Research question one was "What are the experiences of general aviation pilots regarding the integration of TAA?" Table 11 provides a summary of the responses to the question and the supporting questions from the Interview Research Guide.

Table 11 - Commercial-Instrument Pilot Group Responses to Research Question 1 (RQ1)

Category of Response	Frequency	Percentage
Regulatory Change	11	84.6
More Comfortable	8	66.7
Standardized Training	8	66.7
Limit Equipment	6	46.1
Frequent Operations	5	41.6
Training Limitation	5	41.6
Erosion of Skills	3	23.1

n=13

Eight-four percent of the participants recommended that no major regulatory change is necessary for the integration of TAA in general aviation: however, the participants recommended that pilots earn a TAA endorsement from a certified flight instructor after receiving flight instruction in a glass cockpit. CIR-7 stated, "There is no guarantee that pilots are receiving glass cockpit training, so enforcing general aviation pilots to receive a TAA endorsement will help reduce the number of accidents." The participants answered the question to reflect their experiences with TAA. Majority of the participants (66.7%) responded they felt more comfortable when flying TAA. For example, CIR-8 stated, "When flying on cross country flights I used the autopilot and the global positioning systems 95% of the time because I feel safer and less stress in glass cockpit." Additionally, CIR-2 avowed, "Since I transitioned to TAA I will never fly conventional aircraft again because of the safety, reassure, and built in redundancy". Sixty-seven percent of the participants lamented that the lack of standardized training programs degrades the quality of training for pilots. Case in point, CIR-6 stated, "Part 61 Programs suffer the most since this training paradigm lacks structure, I had to which to a Part 141 Program when getting my instrument rating in order to receive decent glass cockpit training." Forty-six percent of the participants reported the overall quality of training and progress of TAA in general aviation was affected by the multiple types and variants of glass cockpits. CIR-9 stated, "During my training I trained in Cessna, Cirrus, and Diamond aircraft that all had different glass cockpit systems; it was confusing and frustrating initially." Forty-one percent of the participants responded when flying TAA that pilots have to fly regularly to stay proficient. For instance, CIR-3 stated, "There was time when I had about 60 days between flights, I struggled in the aircraft due to my skills deteriorating, and I had to hire a certified flight instructor for a training flight." CIR-8 commented, "When flying TAA, your skills become perishable when there are long periods between flights; especially when you are flying solo." Forty-one percent of the participants commented on training limitation due to a shortage of flight training instruction for TAA. CIR-13 stated, "There are too few locations to get quality training. I reside in a rural area and have to travel 90 minutes in order to receive quality flight instruction on TAA." Some participants (23.1%) alluded that flying TAA degraded their basic airmanship skills. Case in point, CIR-5 stated, "I struggle when I fly conventional aircraft because I have to do all the manual calculations, navigation, and avionics and that makes things difficult when one is accustom to flying TAA." CIR-4 responded, "My basic airplane handling skills have suffered since I started flying glass cockpits, because I rely on the automation – I am training to be an airline pilot – and studies have shown that automation diminishes pilots' flying skills."

Research question two was, "How can general aviation pilots use the automation of TAA to reduce accidents?" Table 12 provides a summary of the responses to the question and the supporting questions from the Interview Research Guide. The participants answered the question to reflect their experiences with the automation of TAA. The majority of the participants (76.9%) responded that leveraging the automation of TAA enhance situational awareness.

Table 12 - Commercial-Instrument Pilot Group Responses to Research Question 2 (RQ2)

Category of Response	Frequency	Percentage
Situational Awareness	10	76.9
Training	10	76.9
Established Personal Standards	7	53.8
ADM / SPRM	4	30.7
Reliance	3	23.1

n=13

The participants also noted that technologies should be implemented to help increase situational awareness and improve safety. For example, CIR-13 stated,

> After flying over 100 hours in TAA, I feel safer and more informed when operating TAA, I do not have to do the mental math, which allows me to focus on the airplane and where it is going and manage the information rather than produce the information.

A majority of the participants (76.9%) alluded training is the key to reduce general aviation accidents. Case in point, CIR-4 stated,

> The best way to reduce accidents is to train on a frequent basis or fly more consistently to stay proficient. I fly three to four times a month to maintain my skills because flying TAA takes a more advanced approach to flying.

CIR-10 stated:

Sometimes I fly with other pilots to observe their techniques and cockpit management, because it helps me learn new techniques and reinforce my good practices learned while in training. Talking with pilots and instructors gives me the opportunity to learn about the latest technologies and see the new equipment while observing.

A slight majority of participants (53%) suggested pilots have to understand and respect the dangers associated with flying advanced aircraft and should fly within the scope of the aircraft. For example, CIR-5 stated,

I set personal standard and minimums when I fly, which keep from getting in a dangerous situation or flying beyond my means. Glass cockpits were designed to make flying easier but it requires pilots knowing their limits, my limits are my limits, I use them as safety barriers.

Thirty percent of the participants believe that aeronautical decision-making and single pilot resource management were designed to teach pilots to fly advanced aircraft safely. For instance, CIR-14 avowed:

Most of general aviation accidents are due to pilots making mistakes. My aeronautical decision-making is part of me and I have to make quick and accurate decisions all the time when flying so I am accustom to thinking through my decisions. As a single pilot, you have to learn to rely on your decisions and let the automated aspect of the aircraft be a part of your decision process.

CIR-8 stated,

I am not an experienced pilot when it comes to flying TAA, and my flight instructor recognized I did not have a lot of experience in advanced aircraft; he taught me aeronautical decision-making and single pilot resource management to keep me safe and as a proven process to make me a better pilot.

Twenty-three percent of the participants responded that technologies should be implemented to improve safety and for reliance. Case in point, CIR-3 stated,

I love to hand-fly the aircraft to get a better feel for the aircraft, but during my training the flight instructor repeatedly directed me to use the automation in order to free up my time and cognitive abilities. The instructor told me that using the automation prevents fatigue and help me establish trust that the aircraft will perform as designed.

Research question three was, "How can general aviation pilots reduce the effects of technological innovations to capitalize on the positive attributes of TAA?" Table 13 provides a summary of the responses to the question and the supporting questions from the Interview Research Guide. The participants answered the question to reflect their experiences with the automation of TAA.

*Table 13 - Commercial-Instrument Pilot Group Responses to
Research Question 3 (RQ3)*

Category of Response	Frequency	Percentage
Training	13	100.0
Safety	9	69.2
Pilot-centric	7	53.8
Simplify	6	46.1
Combined Training	3	23.1

n=13

All the participants (100%) avowed that training is the best the approach to acclimate and integrate new technologies into one's flying practices. For example, CIR-6 stated,

> The best way to learn new technologies is to train and practice with technology and determine to the most useful way to leverage the new gadgets. I don't support all the technologies in general aviation because inexperience pilots like me use the automated features way too much and fail to hone our basic flying skills, which is why I use every flight as a training flight so I can become a safer pilot.

Sixty-nine percent of the pilots asserted that technologies should be integrated to improve aviation safety. Case in point, CIR-13 stated,

With all the technologies and changes in aviation, there has to be a 'human factors' concern. General aviation pilots are not provided sponsored training like airline and military pilots, so we gravitate to new technologies to be better pilots without knowing the consequences. Unfortunately, I have do not have the experience to determine the human factors impacts. Since transitioning to a glass cockpit my instrument flying has improved but my basic flying skills have eroded due to my extensive reliance on automation.

CIR-4 stated,

I don't feel enough safety emphasis has been placed on TAA in general aviation because after eight flight hours I was given authorization to rent a TAA from a local club and I was nervous and knew that I was not ready to fly alone, so I went back for an additional six hours of training before flying solo.

Fifty-three percent of the pilots emphasized too much attention was placed on the technology, which was wrong, because the focus needed to be pilot-centric. For instance, CIR-3 stated,

During my training, my instructor constantly paid attention to the equipment rather than how I was progressing during the transition. I have made 11 trips as a single pilot and I was not sure if I progressing satisfactorily or not, I just know that I am not nearly as comfortable in TAA as I am in conventional aircraft.

Forty-six percent of the pilots believe general aviation aircraft should be simple. CIR-7 stated, "TAA are too busy and advanced for basic pilots I rather fly basic aircraft so I can understand what is happening and not let the aircraft lead me." To keep pilots basic skills from eroding, 15% responded that pilots should fly both TAA and conventional aircraft. For example,

CIR-2 stated, "Flying both conventional and TAA are beneficial, I fly both to keep my skills from expiring because I want to become a certified flight instructor, so flying both models kept me honest."

Research question four was, "What is the impact of TAA on aeronautical decision-making skills of general aviation pilots?" Table 14 provides a summary of the responses to the question and the supporting questions from the Interview Research Guide. The participants answered the question to reflect their experiences with the automation of TAA. A slight majority of the participants (53%) avowed that general aviation pilots need solid aeronautical decision-making skills to fly TAA.

Table 14 - Commercial-Instrument Pilot Group Responses to
Research Question 4 (RQ4)

Category of Response	Frequency	Percentage
Safety	7	53.8
Competency	6	46.1
Higher Order of Thinking	3	23.1

n=13

For example, CIR-11 stated, "Aeronautical decision-making keeps me safe and makes me think of logical responses and processes when flying, without have a thorough understanding of aeronautical decision-making I would be a below average pilot." Forty-six percent of the participants stated that FITS, aeronautical decision-making and single pilot resource management increase their competency and helped made them better pilots.

Case in point, CIR-5 stated, "If you are serious about flying and being safe in the air then FITS and SPRM are must haves because they will keep you alert when flying and provide a valuable foundation when operating TAA in hectic situation ways." CIR-12 stated, "The first thing that comes to mind when thinking of aeronautical decision-making is higher order of thinking skills. Advancing to TAA requires higher level of thinking and flying such as the progress pilots undergo when becoming an airline pilot."

Certified Flight Instructor (CFI) Group.

This group of general aviation pilots has attained a Certified Flight Instructor license. Research question one was, "What are the experiences of general aviation pilots regarding the integration of

TAA?" Table 15 provides a summary of the responses

to the question and the supporting questions from the

Interview Research Guide. An obvious majority of the

participants (81%) alluded no regulatory change is

required to address the integration of the TAA in

general aviation.

*Table 15 - Certified Flight Instructor Group Responses to
Research Question 1 (RQ1)*

Category of Response	Frequency	Percentage
Regulatory	9	81.8
Transformative	7	63.6
Human Factors	7	45.4
Degradation	7	45.4
Cost	5	45.4

n=11

For example, CFI-6 stated, "Implementing a

regulatory change to address TAA would be overkill

because most of the accidents and fatalities in

general aviation are linked to pilot error. A regulatory

change could diminish the liberties that general

aviation pilots enjoy." Sixty-three percent of the

participants mentioned their experiences with TAA

have been transformative. Case in point, CFI-8

stated,

Flying TAA demands more cognitive abilities, focus because pilots rely on the automation and do not stay engaged in the safe operation of the aircraft, so pilots have to combat the complacency, and stayed involved in the flight operations. This is a side effect of integrating TAA in general aviation.

Forty-five percent of the participant suggested that human factors, degradation, and cost all negatively affect aviation. CFI-3 stated, "I constantly observed pilots struggle with TAA because the cockpit is interactive and most pilots have to learn to modify their cockpit procedures and scan due to advanced avionics and automation of the cockpit." CFI-12 stated, "The degradation of basic flying skills will always exists until we get handle on the human factors deficiency introduced by TAA." Some participants (45%) reported the rising cost to rent TAA has prevented the further investing in flight training so some pilots risk flying TAA without proper training. For example, CFI-9 stated, "The cost to pursue flight training in TAA is too steep so I am forced to reduce the amount of training I receive."

Research question two was, "How can general aviation pilots use the automation of TAA to reduce accidents?" Table 16 provides a summary of the responses to the question and the supporting questions from the Interview Research Guide. The participants (100%) responded that the increased situational awareness from TAA is an enabler to reduce general aviation accidents.

Table 16 - Certified Flight Instructor Group Responses to Research Question 2 (RQ2)

Category of Response	Frequency	Percentage
Situational Awareness	11	100.0
Cognitive	10	90.9
Dependent	5	45.4

n=11

For example, CIR-9 stated, "Without a doubt TAA increase pilots' situational awareness beyond that of conventional aircraft. Often I feel like an information manager instead of a pilot because of the automation and constant information provided in the cockpit." Ninety percent of the participants indicated that TAA is cognitive demanding because data such as weather, speed, fuel burn, distance, speed, wind, and location are available at all times. For example,

CIR-7 stated, "Because of all the data that are provided, I feel that my cognitive abilities are enhanced which allows me to stay ahead of the aircraft and know exactly how the aircraft is performing so I can make informed decisions." Some participants mentioned pilots become dependent on the automated capabilities of TAA. Case in point, CIR-11 stated, "About 500 feet above the airport I engage the autopilot for a smoother transition to cruise flight, so you can say I over-rely on glass cockpit to make my flying easier."

Research question three was, "How can general aviation pilots reduce the effects of technological innovations to capitalize on the positive attributes of TAA?" Table 17 provides a summary of the responses to the question and the supporting questions from the Interview Research Guide. The participants (100%) responded that maintaining situational awareness enabled pilots to reduce the effects of technological innovations.

*Table 17 - Certified Flight Instructor Group Responses to
Research Question 3 (RQ3)*

Category of Response	Frequency	Percentage
Training	11	100.0
Limit Technology	6	54.5
Safety Awareness	5	45.4
Active Learning	3	27.2

n=11

For example, CFI-2 stated, "Learning the aircraft inside and outside would help pilots fly safely and identify any issues associated with technologies because every innovation will impact existing standards and force pilots to develop new processes." Fifty-four percent of the participants responded that pilots should limit their exposure to advanced aircraft until basic flying skills have been mastered. For instance, CFI-6 stated, "I flew sport pilot aircraft before upgrading to TAA, so I could get use to the digital cockpit layout and improve my cockpit span." Forty-five percent of the participants reported that safety awareness was required to capitalize on new technologies. CFI-3 stated,

As a Certified Flight Instructor, some of my students just fly for the sake of flying without realizing the effect of technology on their basic

practices taught in basic flight training – and
that can be catastrophic. Some inexperienced
pilots gravitated to flying aircraft that were
more complex without understanding the
complicated nature of the systems and this
places those pilots at a higher risk for
accidents. I would much rather pilots learn the
systems first rather than learn how to leverage
the automated capabilities of the aircraft.

Some participants alluded general aviation
pilots should take a more proactive role in learning the
intricate details of their aircraft capabilities. Case in
point, CFI-4 stated,

Before determining which aircraft I wanted to
fly, I studied the pros and cons of each aircraft
and its capabilities to make sure I was going to
get the most out aircraft and not invest in
training I would never use. Unlike airline pilots,
general aviation pilots have to take a more
assertive role in determining which airplane is
more suitable for their flying. I prefer to fly
cross-country, so I am more interested in the

autopilot capabilities than the advanced avionics.

Research question four was "What is the impact of TAA on aeronautical decision-making skills of general aviation pilots?" Table 18 provides a summary of the responses to the question and the supporting questions from the Interview Research Guide. The participants answered the question to reflect their experiences with the automation of TAA.

*Table 18 - Certified Flight Instructor Group Responses to
Research Question 4 (RQ4)*

Category of Response	Frequency	Percentage
SPRM	7	63.6
Cockpit Management	6	54.5
Safety Improvement	3	27.2
Confident	3	27.2

n=11

A slight majority of the participants (63.6%) commented that TAA improved the overall coordination of the flight, eliminated the mental workload, and increased the pilot's overall management of the flight. For example, CFI-10 stated, "TAA have streamlined pilot responsibilities that allows the pilot to prioritized flight responsibilities and better managed the flight. Managing the flight is essential to having a successful flight." A slight majority of the participants (54%) responded that flying TAA increased a pilot's cockpit management abilities. CFI-2 stated, "One benefit of flying TAA is improving your cockpit management skills. Having sound cockpit management skills enable pilots to work through their checklists and follow procedures without worrying maintaining heading and altitude which is done by automating in TAA." Twenty-seven percent of the participants commented that TAA improves aviation safety. For example, CFI-3 stated,

> With the proper training TAA could improve aviation safety because the pilot is more informed and not mentally over tasked like when flying conventional aircraft because the pilot functions as a system manager when leveraging all the automated capabilities of the TAA.

Another 27% commented that TAA improved their confidence and decision-making. "Case in point, CFI-11 stated,

I had one student pilot who, before I started

flying TAA, was always timid and flew locally;

now that he has made a transition to TAA, he

regularly makes 250 nautical miles flights on a

routine basis. He is more comfortable,

confident, and reassured because of the

automated and redundancy of TAA.

*Figure 25 - Diagram showing the face of a true airspeed indicator
typical for a faster single engine aircraft; photo courtesy of Mysid.*

Comparison of the Embedded Units of Analysis.

Table 19 provides the statistical data for each embedded units of analysis. This data was essential to compare and understand the responses provided in Table 5. The themes by each embedded unit of analysis were provided in Table 5.

Table 19 - Embedded Units of Analysis for Comparative Analysis

	Private Pilot Group	Commercial Instrument Group	Certified Flight Instructor
Average Age	31.60	37.60	40.70
Average Years as Pilot	5.72	12.46	14.72
Average of Total Flight Hours	729.30	1,290	2,300.20

The private pilot group themes were: (a) challenging transition, (b) training, (c) situational awareness, (d) regulatory, (e) proficiency and familiarization, and (f) aeronautical decision-making and single pilot resource management. The commercial-instrumented rated group themes were comprised of: (a) regulatory, (b) training, (c) situational awareness, (d) safety, and (e) competency. The CFI group themes included: (a) regulatory, (b) situational awareness, (c) training, and (d) single pilot resource management. All embedded units of analysis listed regulatory as a salient theme even though the private pilot group recommended no regulatory change, the pilots recommended an endorsement for general aviation pilots flying TAA.

The private pilot group was the only group that recommended the endorsement, which contradicts the group's suggestion that no regulatory change was required. This anomaly might be based on the group's 5.72 average years of experience as a pilot. The private pilot and the certified flight instructor groups listed challenging transition or transformative as a major theme for RQ1. One contributing factor for this was private pilots were the least experienced of all the embedded units of analysis. Certified flight instructors are responsible for providing flight instruction to general aviation pilots and through the instructor's lens, the certified flight instructors observed the pilots challenge with the transition to TAA. The private pilot and commercial-instrument rated groups listed training as a major theme, the certified flight instructor group did not list training as a major theme for RQ1, which the group's 14.72 average years as a pilot and average total flight hours of 2300.2 might have contributed to the group's non-selection of training as a major theme. Situational awareness was a theme that all embedded units of analysis selected for RQ2. On RQ3, training was listed a theme for all groups because training remains an essential area of concern for majority of the participants. Limiting innovations was a major theme for RQ3 for the private pilot and certified flight instructor groups. The commercial-instrument rated group did not choose the limiting innovation because members of this group are most comfortable with the intricate workings of the advanced systems as qualified instrument rated pilots. On RQ4, each embedded unit of analysis selected different themes. There is no explanation for the various themes between the groups on RQ4.

Evaluation of Findings

The purpose of this qualitative single descriptive case study was to explore the experiences of general aviation pilots on the integration of TAA and to determine how the automation of TAA can be used to increase aviation safety and improve pilots' decision-making skills. This section includes an evaluation of the findings for 35 general aviation pilots (11 private pilots, 13 commercial-instrument rated pilots, and 11 certified flight instructors). The findings were listed by the participant groups and supplemented by document analysis (technical reports and studies) and the technological determinism theory. Analysis of the data derived from one central research and three supporting research questions that accrued seven salient themes: (a) training, (b) safety, (c) proficiency, competency, and familiarization, (d) transformative and challenging transition, (e) situational awareness, (f) decision-making, single pilot resource management, aeronautical decision-making, and (g) regulation. The minor themes identified were (a) limit innovations and (b) cognition.

Evaluation of findings for Research Question (RQ) 1.

What are the experiences of general aviation pilots regarding the integration of TAA? Research question one focused on understanding experiences of general aviation pilots regarding the integration of

TAA. The responses from the participant groups (private pilots, commercial-instrument rated pilots, and certified flight instructors) and document analysis of technical reports and studies were used to attain empirical data on the integration of TAA in general aviation. The combined data from the participant groups and document analysis on RQ1 contributed to three major themes: (a) regulatory, (b) training, and (c) challenging and transformative transition. The comments below underscore the significance of themes that developed from RQ1. Case in point, PP-9 stated, "I work in the aviation industry and there are extensive discussions on the reluctance to update the Federation Aviation Regulations section that pertains to general aviation to directly address implementation of technologies in general aviation."

CFI-8 stated,

As a flight instructor, I noticed pilots struggling when transitioning from conventional aircraft to TAA, which is impacted by no standardized training example and the different variations in TAA. This makes it difficult for inexperience pilots to grasp the fundamentals of flying TAA.

PP-6 opined,

I started training in TAA and had to transition
back to conventional aircraft during my private
pilot training because TAA were too
complicated and I was not proficient in my
basic flying skills. I eventually transitioned to
TAA after getting my private pilot license.

Regulatory.

The majority of participant groups responded
no regulatory change was needed for the integration
of TAA in general aviation. A majority of the private
pilot group participants lamented the FAA leadership
should mandate general aviation pilots receive an
endorsement to fly TAA after completing the
applicable training. Mandating an endorsement for
general aviation pilots to fly TAA contradicted the
private pilot group response that no regulatory change
was necessary, because mandating pilots to get an
endorsement would require a change to existing

regulations. The disparity in responses could be a lack of understanding of the Federation Aviation Regulations. The lack of understanding is because regulations pertaining to general aviation are generic in nature (Layton, 2012).

Most of the private pilot participants noted an endorsement to fly TAA was necessary for safety reasons because the transition from conventional aircraft to TAA was challenging and transformative. The results revealed the private pilot group had the least number of TAA flight hours (collectively) and experience when compared to the CIR and CFI groups as indicated in Table 19. Supplemented data from document analysis highlighted the NTSB (2010) members recommended to the FAA to incorporate TAA training specifics into aeronautical knowledge requirements. This may be why a regulatory change was not mandated. The findings indicated FITS was designed to reduce the number of general aviation accidents by teaching pilots a combination of single pilot resource management, aeronautical decision-making, and risk management (Franza & Fanjoy, 2012).

Training.

All participant groups stressed for mandatory or improved training to ameliorate aviation safety was a theme derived from RQ1. A majority of the private pilot and commercial-instrument rated groups advocated for the enforcement of FITS because of the lack of standardized training programs, which resulted in general aviation pilots being unprepared to fly TAA. This finding was corroborated by a previous study, given that the integration of TAA has not improved aviation safety (Franza & Fanjoy, 2012) because pilots are instructed to operate glass cockpits using conventional aircraft methodologies, which is an ineffective model due to the vast differences between the conventional aircraft and TAA (NSTB, 2010).

Another issue impacting flight training in general aviation was the omission of critical instruction on specific system knowledge. Existent data from the document analysis indicated that the FAA leadership had not mandated FITS as an approved training curriculum (NTSB, 2010).

Existing data specified that training continued

to lag behind the integration of TAA in general
aviation because the existing conventional practices
are not transferrable to glass cockpits; hence,
increasing the probability of general aviation
accidents (McCracken, 2011). Research by Franz and
Fanjoy (2012) indicated the NTSB, FAA, academia,
and the aviation industry were cognizant to achieve
the increased benefits of automation requires a pilot
to have comprehensive understanding of the systems.
This is further corroborated by a NTSB (2010) report
that indicated FAA inspectors do not have a testing
metric to evaluate the pilots' understanding of system
knowledge that has resulted in pilots having a
cognitive mismatch.

Challenging and Transformative Transition.

Another major theme associated with RQ1 was
the challenging and transformative transition to TAA.
Participants noted that transitioning to glass cockpit
was cognitively demanding, foreign, and difficult.
Participants in the CFI group noted the complexity
involved in transitioning to glass cockpits. The pilots
pointedly noted that to maintain the skills required

frequent flying, because infrequent flying results in atrophy. Some participants avowed the different variations in glass cockpits further complicated transitions, especially for pilots that were initially upgrading to TAA. Participants noted the systems were unintuitive and cognitively demanding. The NTSB (2010) report indicated general aviation pilots are susceptible to cognitive mismatch, because current curriculums do not include training objectives that address specific system knowledge for TAA, which substantiated the finding.

The responses corroborated the existing data that indicated pilots are less comfortable operating TAA even after an extensive period of flight training in glass cockpits because conventional training paradigms were not transferable to pilots operating TAA (NTSB, 2010). This human factors issue was perpetuated by the integration of technology in general aviation and dilapidated training curriculum (NTSB, 2010). Additionally, the GAJSC (2012) substantiated the finding by specifying that transition training was not standardized and resulted in accidents due to pilots not being familiar and proficient with the aircraft.

Evaluation of findings for RQ2.

How can general aviation pilots use the automation of TAA to reduce accidents? Research question two focused on the pilots' experiences on using the automation of TAA to reduce accidents. Participant responses and document analysis of technical reports and studies were used to attain empirical data on leveraging the automation of TAA to reduce aviation accidents. The combined data from the embedded units of analysis and document analysis on RQ2 contributed to three major themes: (a) situational awareness, (b) training, and (c) aeronautical decision-making and single pilot resource management.

Situational Awareness.

All the participants in the private pilot group noted that the automation of TAA increased situational awareness. PP-1 stated, "Using the automation and the autopilot reduces fatigue and allows to pilot to stay mentally alert through all phases of flight." Seventy-six percent of the CIR group noted that leveraging the automated capabilities of TAA improved situational awareness. Participants avowed that technologies should be designed to improve situational awareness to reduce general aviation accidents. CIR-13 stated,

> After flying over 100 hours in TAA, I feel safer and more informed when operating TAA, I do not have to do the mental math, which allows me to focus cognitively on the airplane and where it is going and manage the information rather than produce the information.

The CFI group participants (100%) responded that using the automation of TAA enhanced situational awareness. CIR-9 stated, "Without a doubt, TAA increase pilots' situational awareness beyond that of conventional aircraft. Often I feel like an information manager instead of a pilot because of the automation and constant information provided in the cockpit." The participants' beliefs that TAA ameliorated situational awareness contradicted the finding by NTSB (2010) analysts that glass cockpits did not improve general aviation safety as expected.

Figure 26 - Cockpit of a twinjet flight simulator; photo courtesy of The DJ.

Training.

A majority of the participants indicated training is essential in flying TAA to achieve increased safety to reduce general aviation accidents. PP-2 responded, "The general aviation community needs better training models, especially for those pilots operating complex aircraft that are single-piloted." All of the participants responded that training is the key to learning to leverage the capabilities of TAA to capitalize on the automation. PP-6 stated, "I had to invest in an additional 10 hours of flight training in TAA to fully understand how to learn how to maximize the use of the autopilot and navigational capabilities." PP-10 declared, "I had to repeat several flight lessons in the training program just to comprehend all the advanced automation."

The statements provided by the participants substantiated existing data indicating significant changes to flight training curriculums were necessary for pilots to learn new scan patterns, interpret the data, and managing the redistributed workload, which defies existing practices and processes (McCracken, 2010). The focus on training stems from omission of aeronautical knowledge on aviation tests and the lack of a mandatory training curriculum (NTSB, 2010). The findings indicated the current flight training models are ineffective for instructing pilots transitioning from conventional aircraft to TAA.

Aeronautical decision-making and Single Pilot Resource Management.

Slightly over half of the private pilots (54%) responded that regular training in ADM and SPRM was imperative to reducing TAA accidents. The General Aviation Joint Steering Committee (GAJSC) incorporated a safety enhancement initiative to implement flight safety to focus on aeronautical decision-making to reduce the risks associated with the loss of control in general aviation (GAJSC, 2012). PP-2 responded, "The general aviation community needs better training models, especially for those pilots operating complex aircraft in single-piloted operations." Single Pilot Resource Management is general aviation version of Crew Resource Management but most general aviation pilots are not familiar with it." PP-3 stressed, "Aeronautical decision-making saved my life on numerous occasions after experiencing in-flight emergencies; I relied on my training and trained on a regular basis to stay proficient." PP-7 avowed, "Since transitioning to TAA, my situational awareness and ADM have improved ten-fold. I really don't worry about being task-saturated, the glass cockpit is an enormous help in alleviating the mental math and processing when flying." PP-5 affirmed,

I learned of single pilot resource management through a flying associate, my flight school does not teach it so I took flight lessons from a different school to gain exposure and it made me a different pilot with a methodical process to work through any situation when flying.

PP-12 avowed, "I learned single pilot resource management from day one of my transition to TAA, it was a bit much at the beginning once I put all the pieces together it made me a better pilots." Some participants noted aeronautical decision-making and single pilot resource management were designed to teach pilots to perform better pilots in glass cockpits and complex aircraft. CIR-8 stated,

I am not an experienced pilot when it comes to flying TAA, and my flight instructor recognized I did not have a lot of experience in advanced aircraft, so he taught me aeronautical decision-making and single pilot resource management to keep me safe and as a proven process to make me a better pilot.

According to the findings, participants are confident that aeronautical decision-making and single pilot resource management are vital to improving general aviation safety. Published data indicated that aeronautical decision-making and single pilot resource management are integrated in the FITS curriculum, which had not been mandated the FAA leadership (Halleran & Wiggins, 2010; NTSB, 2010). The GAJSC (2012) recommended the use of web-based resources to assist pilots with transitioning to TAA and to enhance the aeronautical decision-making to overcome training deficiencies. The finding was substantiated by the GAJSC (2012) who recommended for the FAA and the aviation industry to develop a flight-training program that incorporates aeronautical decision-making for reducing the number of loss of control accidents in general aviation.

Evaluation of findings for RQ3.

How can general aviation pilots reduce the effects of technological innovations to capitalize on the positive attributes of TAA? Research question three focused on the pilots' experiences on using the automation of TAA to reduce accidents. The responses from private pilots, commercial-instrument rated pilots, and certified flight instructors and document analysis of technical reports and studies were used to attain empirical data on reducing the effects of technological innovations to capitalize on the positive attributes of TAA. The data from the participant groups on RQ3 contributed to two major themes: (a) training, (b) proficiency and familiarization, and (c) safety.

Training.

All of the participants of the private pilot group (100%) responded that training is the key to learning to leverage the capabilities of TAA to capitalize on the automation. A slight majority of the private pilots

(54%) responded that FITS were essential when undertaking transitioning training.. PP-6 stated, "FITS were the foundational training that taught me how to maximize my aeronautical decision-making skills when piloting TAA." A slight majority of the CIR group (53%) indicated too much attention was placed on the technology. This finding was corroborated by the NTSB (2010) report; however, Franza and Fanjoy (2012) did not substantiate the finding in a previous study. Franza and Fanjoy (2012) reported that general aviation pilots are susceptible to data misinterpretations and cognitive mismatches, preventable through adequate training, maintaining proficiency, and increasing the pilot's understanding of aircraft systems, which indicated researchers are concerned with the impact TAA on general aviation pilots. CFI-3 stated,

> With the proper training could TAA improve aviation safety because the pilot is more informed and not mentally over-tasked like when flying conventional aircraft, because the pilot functions as a system manager when leveraging the automated capabilities of TAA.

Most of the participant groups in this research study indicated training is an essential element for being a safe pilot and reducing general aviation accidents. In a previous study, McCracken (2011) indicated pilots developed their own practices for operating TAA due to a shortage of codified strategies, which resulted in pilots developing poor practices. Cassens et al (2011) indicated that flight instructors were not providing adequate instruction on aeronautical decision-making; therefore, without proper exposure to aeronautical decision-making general aviation pilots remain susceptible to poor pilot decision-making. Published data corroborated the finding that flight training is significant to increasing aviation safety and reducing general aviation accidents even though current flight training paradigms require major improvements.

Proficiency and Familiarization.

Published data revealed pilots struggled and felt less comfortable with the transition from conventional aircraft to TAA even after completing a semester of flight training (McCracken, 2011). The findings indicated that proficiency and familiarization with TAA are fundamental to improving safety and reducing general aviation accidents given that in 2013, 52% of the accidents occurred with an instrument rated pilot onboard, and 23% of the accidents occurred with a certified flight instructor onboard (AOPA, 2013). CIR-9 stated, "During my training I trained in Cessna, Cirrus, and Diamond aircraft that all had different glass cockpit systems, it was confusing and frustrating initially." Forty-one percent of the participants responded that when flying TAA, pilots have to fly regularly to stay proficient. For instance, CIR-3 stated, "There was time when I had about 60 day between flights and I struggled in the aircraft due to my skills deteriorating, I had to hire a certified flight instructor for a training flight". CIR-8 commented, "When flying TAA your skills become perishable when there are long periods between flights – especially when you are flying solo." This finding corroborated the existing data that indicated general aviation pilots should fly frequently to maintain proficiency, especially when the pilot operates multiple aircraft with different glass cockpits, which prevents cognitive mismatch and automation surprise (Franza & Fanjoy, 2012; King, 2011). Data from a previous study indicated general aviation pilots are having difficulty maintaining proficiency due to the increasing cost of flying (Shetty, 2012). The increasing cost of flying supports the finding of the difficulty for general aviation pilots to maintain proficiency and familiarization with TAA.

Existing research indicated general aviation pilots are prone to proficiency errors because the pilots' confidence exceeds their skill levels (Kearns & Sutton, 2013). Previous research indicated that exposing general aviation pilots to their actual skill levels through scenario-based training would help pilots cultivate an accurate understanding of one's skill level (Kearns & Sutton, 2013). Evidence the finding was corroborated is when the GAJSC (2012) mandated a safety enhancement for the FAA and industry to encourage pilots to exhibit manual proficiency in the event of an automation malfunction. The GAJSC (2012) also recommended the FAA promote manual proficiency in publications to propagate the importance of maintaining proficiency.

Safety.

Majority of the pilots asserted that technologies should be integrated to improve aviation safety. CIR-13 expressed this concern,

> With all the technologies and changes in aviation, there has to be a human factors concern. General aviation pilots are not provided sponsored training like airline and military pilots, so we gravitate to new technologies to be better pilots without knowing the consequences. Since transitioning to a glass cockpit my instrument flying has improved but my basic flying skills have eroded due to my extensive reliance on automation.

This concern was supported by CIR-4 who stated,

> I don't feel enough safety emphasis has been placed on TAA in general aviation because after eight flight hours I was given authorization to rent a TAA from a local club and I was

nervous and knew that I was not ready to fly alone, so I went back for an additional six hours of training before flying solo.

Aviation researchers postulated that pilots training in TAA are capable of the passing the minimum test standards; however, the training paradigms do not provide enough in-depth lessons to adequately prepare pilots to operate advanced technologies (Whitehurst & Rantz, 2012). Published studies revealed that in 2012 there were 1,160 general aviation accidents and 334 fatalities, and in 2013, there were 948 accidents and 283 fatalities (AOPA, 2013).

Additionally, personal flight operation was the lead category for the most accidents with 74% in 2012 and 76% in 2013 (AOPA, 2013). The findings revealed that general accidents remain a public safety concern. Research by Garibay and Young (2013) revealed that incorporating airline practices provide the appropriate focus and strategies to reduce the number of general aviation accidents and improve aviation safety (Garibay and Young, 2013). An additional implication is the lack of regulatory oversight perpetuates problems since training and implementation of regulations lag behind integration of new technology, which places general aviation pilots at greater risk due to eroding aviation safety.

Evaluation of findings for RQ4.

What is the impact of TAA on aeronautical decision-making skills of general aviation pilots? Research question four focused on ADM for general aviation pilots. The responses from private pilots, commercial-instrument rated pilots, and certified flight instructors and document analysis of technical reports and studies were used to attain empirical data on aeronautical decision-making. The combined data from the embedded units of analysis and document analysis on RQ4 contributed to three major themes: (a) aeronautical decision-making and single pilot resource management, (b) training, and (c) safety.

Aeronautical Decision-making and Single Pilot Resource Management.

Every research participant in the private pilot

group (100%) responded that TAA increased their situational awareness and decision-making ability. PP-7 avowed, "Since transitioning to TAA my situational awareness and aeronautical decision-making have improved tenfold. I really don't worry about falling behind the aircraft, the glass cockpit is an enormous help in alleviating the mental math and processing when flying." PP-12 expressed a similar belief, "I learned single pilot resource management from day one of my transition to TAA, it was a bit much at the beginning but once I put all the pieces together it made me a better pilot." Forty-six percent of the participants stated FITS, aeronautical decision-making, and single pilot resource management increased their competency and helped made them better pilots. The findings revealed the participants were strong advocates for aeronautical decision-making and single pilot resource management because the training improved their ability to manage the automated aspects of TAA and to fly safer and smarter.

Research by Franza and Fanjoy (2012) indicated FITS was an effective training program to teach pilots to fly TAA because the program teaches

aeronautical decision-making, single pilot resource
management, and risk management through a
scenario-based approach. Previous studies indicated
that FITS were designed to teach pilots to integrate
advanced technologies into their flying through
scenario-based training that focused on aeronautical
decision-making, single pilot resource management,
risk management, and techniques that taught the
pilots how to interpret the new displays (Lindo et al.,
2012). Cassens et al (2012) indicated the FAA
stressed the importance of FITS to increase
aeronautical decision-making and single pilot
resource management and to reduce poor decision-
making that cause accidents, which further
corroborated the finding.

Training.

Majority of the participants noted with the
proper training and use of the automated features that
TAA could improve general aviation safety. The
finding revealed with proper training that TAA would
increase situational awareness, reduce the workload,
and manage the redistribution of functions and tasks

by teaching pilots to use aeronautical decision-making when flying. CFI-4 stated, "I had to work hard to become proficient and achieve a higher level of flying to operate TAA like a professional and confident pilot. I invested in FITS to reap the dividends." CIR-8 expressed a similar belief,

> I had a focused plan when transitioning to TAA because my mentor informed me that TAA were tough to master, so I studied and prepared for every flight to learn the automation and systems to get the most out of the aircraft.

Published data revealed the pilot's workload is not reduced, but redistributed, and training is necessary to handle the redistribution (McCracken, 2011). Mosier et al (2012) asserted pilots are prone to memory failures when operating in stressful situations, which is further complicated when pilots have random interaction with aircraft due to automation; thus, making pilots vulnerable to overlook irregularities or subtle changes. Researchers postulated the training paradigms do not provide enough in-depth lessons to prepare pilots adequately to operate advanced technologies (Whitehurst & Rantz, 2012).

Figure 27 - Upgraded "Glass" C-5M Instrument Panel; photo courtesy of S. Voytek.

Refining general aviation flight training is essential to teach pilots best practices to fly automated aircraft and adapt to the technological centric environment to keep the pilot relevant in the flight management decision loop (Haslbeck et al., 2011). The NTSB report (2010) revealed TAA did not improve general aviation safety as expected. The NTSB (2010) report concluded the generalized training was inadequate to train general aviation pilots to fly TAA due to the increasing complexity of general aviation aircraft. The finding was supported by existing researched, which indicated that general aviation flight programs require major refinements to training pilots to fly automated aircraft in complex environments. The GAJSC (2014) corroborated the finding by indicating the lack of transition training has been listed a major causal factor in general aviation accidents.

Safety.

A majority (69%) of the participants indicated FITS and TAA could improve aviation safety. CIR-5 stated, "If you are serious about flying and being safe in the air then FITS and SPRM are must haves because they will keep you safe when flying and provide an appreciated foundation when operating TAA." CIR-12 stated, "The first thing that comes to mind when thinking of aeronautical decision-making is higher order thinking skills. Upgrading to TAA requires higher level of thinking and flying such as the progress pilots undergo when becoming an airline pilot." Research by Franza and Fanjoy (2012) indicated that FITS were designed purposely to teach pilots to make better decisions by using internal and external resources to include leveraging the aircraft's automated capabilities. The finding is corroborated by the NTSB (2010) report that indicated the general aviation pilots could achieve increased levels of safety by completing FITS training. The data indicated that aeronautical decision-making and single pilot resource management are effective training and processes for pilots that fly TAA because of the foundational procedures, which included aeronautical decision-making and single pilot resource management are taught to pilots to manage the flight by integrating technology into the operations, especially during the critical phases of flight and emergent situations (Halleran & Wiggins, 2010; King, 2011). Published data indicated that as the next generation of technologies is integrated into aircraft, designers and industry leaders have to develop holistic training programs to prevent negative effects like automation surprise, automation misuse, and automation overload (Geiselman, Johnson, and Buck, 2013; King, 2011). The finding was partially corroborated because without improved regulatory reform, training will continue to linger behind the continuous integration of new technologies (Downer, 2012).

Findings in the context of the theoretical framework.

The theoretical foundation of this study was based on the technological determinism theory. The basis of technological determinism is the creation and implementation of new technologies improve human lives by enhancing processes and procedures (Chandler, 2012), regardless of the cultural and societal affects (Yang, 2009). This contributed study to the field of study by providing qualitative data regarding the impact of technological determinism on the integration of TAA in general aviation. This study highlighted several disconnects and negative consequences extending from the integration of TAA in general aviation. The findings of this study do not support the technological determinism theory due to the negative impacts on: (a) flight training, (b) safety, (c) proficiency and (d) regulatory oversight, which all aftereffects of integrating TAA in general aviation.

The integration of TAA introduced the abovementioned negative impacts, which did not improve general aviation. A slight majority (54%) of the participants stressed the need of mandating FITS, which will require a regulatory change; consequently, without a regulatory change there will not be any changes to the flight-training model and aviation certifications (endorsement for TAA). The study revealed improved flight training is essential to ameliorating pilots' decision-making and effectively using automation. A major finding was the current flight training models are ineffective for instructing pilots transitioning from conventional aircraft to TAA, because pilots were unqualified to attain increased levels of safety provided by glass cockpits. Additionally, this study revealed that flight-training practices designed for conventional aircraft are not applicable to TAA pilots because the training lacks instructions on system knowledge, aeronautical decision-making, single pilot resource management, and leveraging automation to improve situational awareness.

This study revealed the lack of a comprehensive training program such as FITS highlighted the regulatory disjointedness for not mandating FITS. To achieve the anticipated levels of increased safety, this study indicated the FAA leadership should mandate FITS. Another finding from this study was the reluctance to mandate FITS as the standardized training model prevented pilots from improving their situational awareness, decision-making, risk assessment, task assessment, automation management, and higher levels of safety, which are the pillars of aeronautical decision-making and single pilot resource management. One finding revealed aeronautical decision-making and single pilot resource management are imperative to reducing accidents, because FITS teach pilots higher order thinking skills. Another related finding was general aviation pilots underwent a challenging transition and failed to achieve proficiency, which increased the chances for cognitive mismatch and automation surprise.

Finally, academia, the aviation industry, and government organizations have vital responsibilities in progressing acuity and application of technological determinism in general aviation. This study revealed technological determinism and its applicability to general aviation requires fundamental changes to avoid unforeseeable consequences on flight training, safety, proficiency and regulations. Based on the findings of this study, technological deterministic thinking is one-dimensional and fails to corroborate the impact of technology from a multi-disciplinary approach, such as the aftereffects from implementing technologies that failed to improve aviation safety, reduce poor decision-making, and endangering public safety. This study proved focusing on the technological advantages and disregarding the adverse implications perpetuates the misconception that technology improves society.

Summary

The purpose of this qualitative single descriptive case study was to explore the experiences of general aviation pilots on the integration of TAA and to determine how the automation of TAA can be used to increase aviation safety and improve pilot decision-making skills. Three embedded units of analysis consisting of 35 pilots (11 private pilots, 13 commercial-instrument rated pilots, and 11 certified flight instructors) were interviewed to provide enriched data on the integration of TAA in general aviation and ways to use the automation to improve aviation safety and decision-making. Document analysis performed on technical reports and studies supplemented the interview data. This chapter included the findings, which were broken down by the primary research question and sub-questions. The findings were then interpreted in the evaluation of findings sections, which reviewed seven major themes identified in the analysis of the data. The explanation of the results of the study was based on the study's theoretical frameworks and the findings were compared to the supplemented data and findings of previous scholarly research in the field.

Evaluations of findings revealed seven theme salient themes. With respect to the integration of TAA in general aviation and the impact on aviation safety and pilots' decision-making the following themes emerged: (a) training, (b) safety, (c) proficiency, competency, and familiarization, (d) transformative and challenging transition, (e) situational awareness,

(f) decision-making, single pilot resource management and aeronautical, and (g) regulation.

The study contributed to technological determinism by exposing that technological deterministic thinking does not account for after-effects of integrating TAA in general aviation. The study revealed the integration of technology in general aviation did not improve existing processes and institutional practices, in fact, flight training, safety, and regulations were highlighted as areas that require immediate attention and action. This study can be used to provide the framework to expand the technological determinism theory to account for the aftereffects of new technologies by developing a model for multi-disciplinary domains like aviation.

When compared to other studies, this study produced analogous results. The comparison revealed disparities and similarities between different participant groups regarding the integration of TAA in general aviation and determining the impact of automation on aviation safety and pilots' decision-making. In Chapter 5, the researcher discussed implications of findings and recommendations for future studies.

CHAPTER V
SUMMARY, CONCLUSIONS, AND
RECOMMENDATIONS

The purpose of this qualitative single
descriptive case study was to explore the experiences
of general aviation pilots on the integration of TAA
and to determine how the automation of TAA can be
used to increase aviation safety and improve pilot
decision-making skills. A significant problem with
integrating TAA in general aviation is the lack of a
systematic approach for integrating TAA that
addresses the impacts on current system designs,
processes, and regulations (Garibay & Young, 2013;
Pritchett, 2009; Robertson, 2010). The general
aviation sector needs a systematic process to
integrate technologies without introducing new
hazards and safety concerns (Di Renzo, 2010). This
qualitative single descriptive case study was
conducted using semi-structured interviews, notes
from the interviews, data from the pilots' flight

logbooks, and technical studies and reports to explore pilots' experiences on the integration of TAA. To gain comprehensive perspectives, this case study explored a single case with three embedded units of analysis or participants groups.

A major limitation of this study was restricted to general aviation pilots in Georgia, South Carolina, and Florida, which was a small representation of the general aviation pilot population. Given the scope and nature of the study, restricting the study to general aviation pilots was a limitation because corporate, military, and airline pilots have significant experience with flying TAA. Another limitation was getting the research participants to speak freely and honestly when responding to the research questions. To gain an in-depth perspective on the integration of TAA in general aviation required the research participants to provide unencumbered information on their personal experiences.

Thirty-five general aviation pilots were interviewed for this study. The general aviation pilots were private pilots, commercial-instrument rated pilots, and certified flight instructors, which represented the three participant groups and the three embedded units of analysis. The collected data from the semi-structured interviews and the supplemented data (technical studies and reports) explored the impact of integrating TAA in general aviation and determining how to use automation to improve aviation safety and pilots' decision-making. The technical reports and studies assisted with triangulating the information with the interview data, which increased the reliability and accuracy of the findings. This research was accomplished using a qualitative descriptive single case study methodology. The single case study approach was strengthened by using three participant groups to provide accounts on personal experiences with TAA based on different flight qualifications and a comparison of the findings. This single case study lacked the inductive reasoning of a multiple case study; therefore, triangulation was used to increase reliability and accuracy of the findings and to reduce threats.

The ethical factors of the study focused on safeguarding the research participants' anonymity throughout the study and protecting the participants from harm while assuring truthfulness, reliability, and competence. Participants were provided assurances that partaking was voluntary and pseudonyms were used to conceal the participants' identities. The researcher ensured the research participants were not exposed to any danger or maltreatment during the study. All research participants signed an informed consent statement before participating in the study. The research participants were educated on the confidentiality and anonymity practices as a measure to protect their identities to acquire credible, factual, and reliable data during the semi-structured interviews. Therefore, it was essential to establish an open rapport with the research participants. Chapter 5 includes the study implications, recommendations concerning the relevance of the findings, recommendations for future research, and concludes with a summary of the prominent details.

Implications

The results of this study derived enriched data from one central research question and three sub-questions and content analysis from technical reports and studies. This case study produced the following major themes: (a) training, (b) safety, (c) proficiency, competency, and familiarization, (d) transformative

and challenging transition, (e) situational awareness,

(f) aeronautical decision-making and single pilot

resource management, and (g) regulation. The

inferences of the findings and their applicability to the

themes are listed under the respective research

questions.

Implications based on the findings of RQ1.

What are the experiences of general aviation

pilots regarding the integration of TAA? Three

implications related to RQ1were observed from

evaluating the findings: (a) regulatory amendments,

(b) improved flight training, and (c) challenging

transition from conventional aircraft.

Regarding general aviation pilots' experiences

with the integration of TAA, majority of the participants

felt that no regulatory change was necessary to

govern the integration of TAA in general aviation.

Most of the private pilot group asserted the FAA

leadership should mandate for general aviation pilots

to get an endorsement to satisfy training requirements

for TAA. Members of the NTSB reported training in

conventional aircraft did not prepare general aviation pilots to fly TAA (Franza & Fanjoy, 2012). The lack of preparation by general aviation pilots impelled the private pilot group to advocate for a TAA training endorsement. Previous research indicated the integration of TAA in general aviation was problematic and indicative of the unique challenges of leveraging new technologies to improve safety (Lindo et al., 2012). The problems associated with TAA in general aviation are: (a) out-dated flight training practices, (b) the nonexistence of a government and non-government committee to regulate technologies in general aviation, (c) no defined requirements for TAA proficiency training, and (d) the lack of a requirement to instruct general aviation pilots on single pilot resource management (Franza & Fanjoy, 2012). Therefore, due to personal experiences, majority of the private pilot group recommended mandating a training endorsement as a measure to increase aviation safety and to improve TAA flight training curriculums.

The disparity between the participant groups was indicative of flight experience and flight hours in TAA. A finding in a previous study indicated general aviation accidents occur mostly between 50 and 350 flight hours (Knecht, 2013). The CIR and CFI groups had more flight experience and flight hours in TAA than the private pilot group, which led to differences in the responses between the CIR and CFI groups and the private pilot group. The implication of mandating an endorsement to fly TAA would require a regulatory change to the Federal Aviation Regulations (FAR). Another reason for the differences between the groups is a lack of understanding of the FAR by general aviation pilots. In a previous study, general pilots expressed concern with the complexity surrounding regulations and the number of organizations involved and affected by changing regulations (Shetty, 2012). Without implementing changes to the FAR, the general aviation community will continue to be plagued by new technologies due to ineffective strategies and practices that do not improve aviation safety and pilots' decision-making skills. Published data indicated that aviation safety and pilots' decision-making were reduced due to a lack of familiarization with technology and the complexity of automated systems that resulted automation misuse (Dobson, 2012). Implementing a change to FAR to mandate FITS might eradicate such mistakes and improve aviation safety.

Participants in this study indicated that FITS was an effective program to train pilots transitioning from conventional aircraft to TAA. The findings of the study indicated participants recommended the development of standardized flight training program. Participants indicated the transition from conventional aircraft was difficult, challenging, and convoluted. The difficulty associated with general aviation pilots transitioning to TAA were perpetrated by the lack of an effective training program (McCracken, 2011). According to King (2011), the FAA leadership and the aviation industry recognized the need for FITS but FITS remained a recommended and highly endorsed program rather than mandated training. Existing scholarly research does not indicate why the FAA has not mandated FITS. King (2011) suggested the FAA leadership relies on aircraft manufacturers and commercial vendors to provide regulated-level glass cockpit training.

Without FITS being a FAA-mandated program, general aviation pilots would not receive the comprehensive training on the intricate workings of the automation (King, 2011). The participants advocated for FITS due to their personal experiences and aspiration to have an in-depth training program, which would improve their decision-making and improve aviation safety. Participants noted the aircraft avionics and systems were unintuitive and cognitively demanding, which supported King's (2011) argument that the training provided by aircraft manufacturers and commercial vendors failed to provide comprehensive instructions on aircraft systems. The participants provided ample comments and suggestions indicating that existing training paradigms were ineffective and endangered general aviation safety. Another noticeable practice in general aviation was that pilots were expected to gain operational skills and proficiency through experience (Drinkwater & Molesworth, 2010) even though existing flight training paradigms do not prepare general aviation pilots to fly TAA. In addition, this finding supported a similar conclusion by McCracken (2011) that additional changes are needed regarding the transition from conventional aircraft to TAA for general aviation pilots.

This study highlighted the request from general aviation pilots to mandate FITS as the standardized flight-training model for TAA. Considering the specific problem of this study, to determine the impact of TAA on general aviation and the use automation to improve aviation safety and pilots' decision-making abilities, this study suggested the general aviation population aspired to mandate FITS and a flight training endorsement to fly TAA. This finding contributed to the existing body of knowledge by specifying the aftereffects from implemented technologies require commensurate focus and attention that led to innovation implementation. The implication of not understanding the training and human factors aftereffects perpetuated pilots' poor decision-making, which resulted from automation overload, automation misuse, and substandard training.

Implications based on the findings of RQ2.

How can general aviation pilots use the automation of TAA to reduce accidents? Three key implications were associated with RQ2 as a result of the findings: (a) situational awareness, (b) training, and (c) aeronautical decision-making and single pilot resource management.

Regarding the use of the automation of TAA to reduce accidents, majority of the participants indicated that the automation could be used to increase situational awareness because the advanced capabilities provided readily available and real-time data to the pilot. A NTSB (2010) report indicated TAA did not improve general aviation safety. The participants in this study suggested that TAA improved their situational awareness and decision-making. The implication is the study participants acknowledged the increased safety, which contradicted indications from the NTSB (2010) report on TAA. The contradiction highlights opposing perspectives between what general aviation pilots are currently experiencing and what previous studies indicated about general aviation pilots' experiences in TAA (McCracken, 2011; NTSB, 2010). One reason for the contrasting views was the general aviation community failed to capitalize on free educational and training resources; thus, abridging strategic efforts by the NTSB and FAA to improve general aviation safety

(Garibay & Young, 2013).

Additionally, the participants indicated better flight training was required in order to maximize the use of automated capabilities to reduce accidents. This finding corroborated a previous study that specified general aviation pilots had routinely misinterpreted data provided by automated technologies (Vincent et al., 2014). McCracken (2011) and Vincent et al (2014) specified that extensive changes to current flight training paradigms are needed to teach general aviation pilots to correctly interpret and capitalize on the advanced capabilities of TAA, which was corroborated by the participants in this study. McCracken avowed that TAA redistributed the workload, which required general aviation pilots to undergo training like FITS to handle the reallocation of tasks. The participants and supplemented data indicated that aeronautical decision-making and single pilot resource management are effective processes and tools for general aviation pilots that operating TAA because both teach higher order thinking skills in order to make better decisions. This finding highlighted a limitation of technological determinism given that the aftereffects were due to the integration of technology.

Figure 28 - The cockpit of the Concorde, which has an 'M'-shaped yoke mounted on a control column; photo courtesy of Christian Kath.

The finding indicated the participants were interested in leveraging automation and technological capabilities of TAA to reduce accidents. This finding was corroborated in a previous study that indicated general aviation pilots are 14 times more likely to be involved in a fatal crash because of flying in inclement weather, single-pilot operations, extensive multitasking during departure and approach phases, and communicating with air traffic controllers (Shao, Guindani, & Boyd, 2014). General aviation pilots are interested in leveraging the automation of TAA to reduce the workload by offsetting some of the tasks during operational intensive phases. A report by the Government Accountability Office (2012) specified the leadership of the FAA was not committed to using technologies to improve general aviation safety, because most of general aviation accidents were due to pilot error. This highlights diverging perspectives between the FAA and general aviation pilots on determining the appropriate innovations to improve aviation safety without introducing unintended consequences or aftereffects. In fact, the aviation community invested over 2.3 million dollars to develop FITS; yet, the FAA leadership declined to mandate FITS as the standardized benchmark for TAA (Halleran & Wiggins, 2010); therefore, FITS remains recommended training for general aviation pilots. According to a NTSB (2010) report, TAA has not improved general aviation safety as expected, which is evident by the NTSB investigation of 1,600 general aviation accidents annually (Garibay & Young, 2013). A previous study indicated a decline in general aviation operations coupled with pilots' pessimism for expanding regulations, and increasing cost of general aviation flying were challenges of the aviation industry (Shetty, 2012), which is a conceivable reason why the FAA leadership has not implemented FITS. Researchers indicated the FAA opted for a non-regulatory approach to reduce general aviation accidents through an assertive, data-oriented initiative to decrease accidents by 1% each year (Garibay & Young, 2013).

The participants expressed that ADM and SPRM are essential in improving general aviation safety; therefore, implementing TAA as a training addendum are necessary to equip pilots with the required skills to operate TAA. The increasing complexity of new technologies require flight training curriculums to undergo continuous upgrades, which makes it difficult to develop and implement flight training programs to keep pace with the expanding technological capabilities (King, 2011). Participants suggested training is essential to maximizing the capabilities of TAA to improve situational awareness to reduce general aviation accidents. General aviation accidents remain rampant (AOPA, 2013) even though technologies were implemented to improve aviation safety at an unprecedented rate, which is indicative of ineffective flight training models and practices (King, 2011; McCracken, 2012). The persistent impacts of ineffective training were because of the omission of aeronautical testing knowledge on aviation tests accompanied by the lack of a mandatory training curriculum (NTSB, 2010). For pilots to capitalize on the technological capabilities of TAA to reduce accidents, there is a need to an emphasis improving pilots' decision-making skills as indicated by research participants' consistent recommendations for improved flight training.

In addition, the finding on how general aviation pilots can use automation to reduce accidents revealed the extensive use of conventional aircraft flight training practices for pilots transitioning to TAA did not properly prepare pilots to operate glass cockpits. Previous studies (Franza & Fanjoy, 2013; McCracken, 2012) reported pilots were improperly trained to operate TAA. For pilots to use automation to reduce accidents, there is a need to require pilots to have an in-depth understanding of the advanced avionics and automated systems (Franza & Fanjoy, 2013). Members of the NTSB (2010) indicated for pilots to use automation to reduce accidents, the pilots must have comprehensive equipment-specific knowledge to become proficient at flying TAA. This is essential because a previous study (Franza & Fanjoy, 2013) indicated that a majority of the accidents in TAA were single-piloted. Previous results coupled with this study's findings, highlight the lack of single pilot resource management focused training. This implies a comprehensive program such as FITS is an appropriate training program, because single pilot resource management is an integrated area of focus in FITS. As observed in this study and previous studies (AOPA, 2013; Franza & Fanjoy, 2012; McCracken, 2012; NTSB, 2010), TAA are problematic in general aviation; for example, some scholars suggested that new technologies distract pilots from performing routine tasks (Halleran & Wiggins, 2010). The participants recognized the need for improved flight training; however, the general aviation community struggles to keep pace with the technological innovations as indicative of the dilapidating flight training models.

Within the context of technological determinism, technological advances in aviation have outpaced the general aviation community's ability to develop and implement updated flight training. As observed in this study and previous studies (AOPA, 2013; Franza & Fanjoy, 2012; McCracken, 2012; NTSB, 2010), TAA are problematic in general aviation; for example, some scholars suggested new technologies distract pilots from performing routine tasks (Halleran & Wiggins, 2010). The participants recognized the need for improved flight training; however, the general aviation community struggles to keep pace with the technological innovations as indicative of the dilapidating flight training models.

Implications based on the findings of RQ3 and RQ4.

RQ3 and RQ4 were combined due to the similarity of the implications. How can general aviation pilots reduce the effects of technological innovations to capitalize on the positive attributes of TAA? What is the impact of TAA on aeronautical decision-making skills of general aviation pilots? Four key implications were associated with RQ3 and RQ4 as a result of the findings: (a) training, (b) proficiency, (c) aeronautical decision-making and single pilot resource management, and (d) safety.

Most of the participants suggested flight training was the key to leveraging the capabilities of TAA to capitalize on the automation. Interestingly, the participants declared significant focus was on the technologies rather than the technological impact on the pilots. Previous research indicated the continuous integration of technological innovations influenced the general aviation sector by requiring new flight training requirements such as teaching pilots to manage flight information rather than focusing on the development of technical skills (Robertson, 2010). This finding supported analogous findings (Parasuraman and Manzey, 2013) that automation modified human interaction rather than eliminated tasks as designed, which led to unanticipated and unintended consequences that resulted in the misuse or disuse of automated capabilities. The desired outcomes of improving aviation safety and increased efficiency were not always achieved; consequently, resulting in a pilot training deficiency due to improper preparation in using automated capabilities (Parasuraman & Manzey, 2013). Published data indicated the rapid integration of technologies in the aviation industry threatened existing safety practices due to scarcity of empirical data regarding the impact of technologies on aviation (DiRenzo, 2009; Lee, Jeon, & Choi, 2012). Government authorities have not mandated predictive analysis, which obliges human factors experts to evaluate the adverse impacts of emerging technologies before widespread integration (Berry, Stringfellow, & Shappell, 2009). This highlighted a limitation with technological deterministic thinking. Additionally, this finding underscored the scholastic oversight on the impact of technologies on general aviation pilots.

The findings indicated participants advocated for FITS as an effective foundation and scenario based training to understand how to operate TAA. This is indicative that FITS training is advantageous and needs to be mandatory for general aviation pilots that fly TAA. In fact, FITS was designed by academia and the aviation industry to provide comprehensive instruction in: (a) risk management, (b) aeronautical decision-making, (c) situational awareness, and (d) single pilot resource management (Franza and Fanjoy, 2013; Halleran & Wiggins, 2010). The implication of these findings highlighted the FAA non-

regulatory approach regarding FITS (Halleran &
Wiggins, 2010). The FAA does not intend to mandate
FITS because of its non-regulatory stance on FITS;
however, the FAA continues to work with the aviation
industry and commercial vendors to develop and
improve FITS (Halleran & Wiggins, 2010; King, 2011).
Research indicated some of the innovations
introduced consequences that threatened pilots'
workload, diminished situational awareness, eroded
basic flying skills, and incompatibility between aircraft
and air traffic control systems (Archer, 2012; Doorn,
2012; McCloy, Harper-Sciarini, Durso, Jentsch, Kanki,
& Rogers, 2011; Oster, Strong, & Zorn, 2013). FITS
were designed by the aviation industry, academia,
and the FAA to meet regulatory standards (King,
2011); however, some researchers indicated the
reluctance by the FAA to mandate FITS is a
regulatory oversight because the aviation industry
relies on technology to improve aviation safety
(Halleran & Wiggins, 2010; Oster, Strong, & Zorn,
2013).

In addition, the findings in this case study
indicated proficiency and familiarization were
essential to general aviation pilots transitioning to

TAA and improving aviation safety. This study indicated there was a direct linkage between training and safety because the two factors were conversely related. A previous study indicated general aviation pilots undergo a proficiency check every 24 months coupled with the fact that most flights in TAA are single-piloted (Garibay & Young, 2013). A major concern for maintaining proficiency is the complexity and number of resources to manage in TAA rivals those in commercial and military aviation (Halleran & Wiggins, 2010) accompanied by the fact that most TAA operations are single-piloted. The concern with single-pilot operations in general aviation is the redistributed workload of TAA (McCracken, 2011) and the increasing complexity of technologies that make pilots prone to automation surprise, automation misuse, and automation overload (Geiselman, Johnson, and Buck, 2013; King, 2011). The significance of this finding was to emphasize that better proficiency standards in single-piloted TAA operations are needed. This conclusion supported the finding that better flight training is required for pilots that are flying TAA: (a) to prevent accidents, (b) to improve aviation safety, (c) to capitalize on free training opportunities, and (d) to teach pilots higher order thinking skills. The findings underscore the limitation of technological deterministic thinking, because innovations are to be managed and controlled in order to lessen the aftereffects such as: (a) aviation hazards, (b) improperly trained pilots, and (c) the lack of mandatory training.

Figure 29 - A Lufthansa Flight Training Bonanza 33 at Phoenix Goodyear Airport USA; photo courtesy of Hinty.

The conclusions of this study support findings of previous studies (Garibay & Young, 2013; Halleran & Wiggins, 2010; NTSB, 2010) that specified FITS as the most readily available and thorough flight training for general aviation pilots flying TAA, which better prepare pilots and improve general aviation safety.

The FAA invested in the development of FITS but has not mandated the training; therefore, some researchers indicated the FAA leadership relies on non-governmental entities to increase effectiveness of FITS. The FAA has opted for a non-regulatory approach to reduce general aviation accidents through an assertive, data-oriented initiative (Garibay & Young, 2013). Some researchers argued the lack of regulatory oversight perpetuates the problem since the integration of new technologies created unintended consequences that are only rectifiable through the implementation of new regulations (Garibay & Young, 2013; Lindo et al., 2012; USDOT, 2012).

RQ1: Interpretation.

Findings substantiated the disparity between the participants groups on the need for regulatory changes and the FAA's non-regulatory approach to improving aviation safety. To mandate FITS would require a regulatory change; thus, the FAA has opted for a non-regulatory approach to reduce general aviation accidents. Results further indicated general aviation pilots are hesitant to advocate for regulatory changes due to the complexity and convoluted process of implementing new regulations. Findings indicated that FITS were widely recommended by the research participants as a training program that would improve aviation safety, ameliorate general aviation pilots' decision-making abilities, and provide the in-depth level of training to acquire higher order thinking skills. Findings denoted existing flight training practices are ineffective and did not properly train general aviation pilots to fly TAA. An additional finding specified for better transition training for general aviation pilots upgrading to the TAA from conventional aircraft. The transition was challenging due to outdated flight training curriculums and the lack of instruction on aircraft systems and familiarization.

RQ2: Interpretation.

The results suggested that for general aviation pilots to leverage the automation of TAA to reduce accidents require the pilots to undergo comprehensive training such as FITS, which would increase the pilots' situational awareness, risk management, automation management, and mental capacity. To achieve the increased safety levels, improved flight training curriculums such as FITS are needed to teach pilots aeronautical decision-making, single pilot resource management, and risk management. The results specified that FITS are designed to teach general aviation pilots to leverage automation in an effective manner without inducing automation misuse and automation overload.

Existing flight training modules are tailored primarily for conventional aircraft rather than TAA of because a non-regulatory approach to improve general aviation safety. Failure to change the existing flight training curriculums will prolong efforts to reduce the number of general aviation accidents. Therefore, general aviation pilots need a comprehensive training curriculum that provides pilots with advanced skills and processes to maximize the use of automated aircraft to reduce general aviation accidents.

RQ3 and RQ4: Interpretation.

RQ3 and RQ4 interpretations were combined due to analogous findings. Similar to RQ2, the results suggested that improved flight training was the key to leveraging the capabilities of TAA to capitalize on the automation. Given the participants' responses, the rapid integration of innovations in general aviation continues to plague pilots because changes to flight training models continue to lag behind the implementation of new technologies.

The results suggested better flight training to: (a) reduce accidents, (b) improve aviation safety, (c) capitalize on free training opportunities, and (d) teach pilots higher order thinking skills. Additionally, participants stressed the difficulty in pilots transitioning to TAA because of ineffective flight training; therefore, the participants recommended changes to existing proficiency flight checks. The proficiency issue and flight training problem exist due to inconsistencies and a non-regulatory approach regarding the integration of TAA in general aviation.

Recommendations

The results of this study embody a significant contribution to scholarly research on general aviation and technological determinism by highlighting the aftereffects of integrating TAA and the impacts on using automation to improve aviation safety and pilots' decision-making. This case study concluded that: (a) general aviation pilots are unprepared to fly TAA, (b) existing flight training paradigms are ineffective, (c) the integration of TAA in general did not improve aviation safety, (d) without mandating FITS general aviation will continue to be impacted by aftereffects, and (e) FITS will improve aviation safety and pilots' decision-making skills. In spite of the underlying limitations, the below recommendations were based on the results of the findings, analysis, and conclusions of the study.

Recommendations for practical applications.

The findings of this study highlighted the impacts of technological deterministic thinking on general aviation. This study has practical applicability to the aviation industry, academia, and general aviation. Based on the study findings, improved flight training is required to ameliorate aviation safety by mandating FITS for general aviation pilots. The implementation of FITS would better prepare pilots to operate TAA while improving pilots' decision-making skills and aviation safety. The following recommendations are necessary to combat the aftereffects of integrating TAA in general aviation: (a) continue the strategic efforts between the aviation industry and commercial vendors to improve FITS, (b) leverage scholarly research to justify the importance of FITS, (c) evaluate accidents involving TAA to determine if technology contributed to the accident, (d) encourage pilots to capitalize on free government-sponsored education and training, and (e) increase data collection efforts to determine the negative attributes of integrated technology. These recommendations are aimed towards improving aviation safety, flight training, and using automation to enhance pilots' decision-making skills. These recommendations might increase the practical understanding of technological determinism.

Future Research.

Additional research related to the integration of TAA in general aviation is recommended. The first recommendation is to conduct future research on the

impact of FITS on general aviation pilots to determine
the effectiveness of FITS on improving aviation safety
and pilots' decision-making skills. The recommended
study is necessary because the aviation industry and
the FAA leadership are advocates of FITS; yet, there
is a shortage of empirical data on FITS. The research
could provide empirical data and findings to
substantiate the effectiveness of FITS and its impact
on aviation safety in general aviation. The second
recommendation is to conduct research on the FAA
non-regulatory approach to reduce general aviation
accidents through a data-oriented strategy (Garibay &
Young, 2013). It is imperative to determine the
effectiveness of the non-regulatory approach and its
impact on general aviation safety. The study might
yield scholarly data that influence changes to
regulations to improve general aviation safety due to
the reluctance to change regulations regarding TAA
and FITS. The third recommendation is to conduct
additional research on a systematic approach for
integrating TAA or technologies that account for the
aftereffects on current system designs, processes,
and regulations. This might lead to a systematic
process that identifies problems or issues that impede

safety, regulations, and flight training deficiencies before widespread implementation of new technologies.

Conclusions

The purpose of the this qualitative case study was to explore the experiences of general aviation pilots on the integration of TAA and to determine how the automation of TAA can be used to increase aviation safety and improve pilot decision-making skills. The discourse on the results of this qualitative single case study, the implications, and recommendations are included in this chapter. The problems articulated in this qualitative single case study are centric to the integration of TAA and the impact on aviation safety and pilots' decision-making skills. The integration of TAA in general aviation has resulted in pilots being unprepared to fly TAA, outdated flight training curriculums, and the lack of mandatory training, which have resulted in TAA not improving aviation safety and pilots' decision-making skills. The case study examined the implementation of TAA through the lens of technological determinism to explore how general aviation pilots can use the automation of TAA to improve aviation safety and pilots' decision-making skills.

This study had one central research question and three sub-questions, which focused on personal experiences with technologically advanced aircraft, aeronautical decision-making, technological innovations, and aviation safety. The conclusions from this study were based on the thematic analysis of the semi-structured interviews and document analysis of supplemented data (technical report and studies). The implications of the study listed in this chapter were based on the central research question and three sub-questions. A summary of the findings for this study are as follows: (a) regulatory changes, (b) improved flight training, (c) challenging transition from conventional aircraft, (d) more focus on proficiency and familiarization, and (e) incorporating aeronautical decision-making and single pilot resource management training to improve pilots' decision-making and aviation safety.

The recommendations presented in this study can be used to improve the integration of TAA in general aviation. The practical recommendations are

(a) continuing strategic efforts to improve FITS, (b)

leveraging pilot organizations and academia entities

to advocate for FITS and improved aviation safety, (c)

implementing data collection to evaluate causal

factors in TAA related accidents, and (d) encouraging

pilots to capitalize on free government-sponsored

educational and training opportunities. The three

recommendations for future research involved

determining the impact of FITS on general aviation

pilots to evaluate if FITS improved aviation safety and

pilots' decision-making, and to conduct future

research on the how TAA influence human factors in

general aviation pilots.

REFERENCES

Aircraft Owners and Pilots Association Air Safety Foundation. (2010).
2010 Nall Report. Accident trends and factors for 2009. Retrieved
from http://www.aopa.org/asf/publications/09nall.pdf

Aircraft Owners and Pilots Association Air Safety Foundation. (2011).
2011 Nall Report. Accident trends and factors for 2010. Retrieved
from http://www.aopa.org/asf/publications/10nall.pdf

Aircraft Owners and Pilots Association Air Safety Foundation. (2012).
2012 Nall Report. Accident trends and factors for 2011 Retrieved
from http://www.aopa.org/asf/publications/11nall.pdf

Aircraft Owners and Pilots Association Air Safety Foundation. (2013).
2013 Nall Report. Accident trends and factors for 2012. Retrieved
from http://www.aopa.org/asf/publications/12nall.pdf

Allen, K. (2012). I hope: A phenomenological study of the lived
experience of fragile families. Marriage and Family Review, 48(7),
621-642. doi:10.1080/01494929.2012.691084

Anderson, C. (2010). Presenting and evaluating qualitative research.
American Journal of Pharmaceutical Education, 74(8), 1-7.
doi:10.5688/aj7408141

Archer, J. (2012). Effects of Automation in the Aircraft Cockpit
Environment: Skill Degradation, Situation Awareness, Workload
(Doctoral dissertation, School of Industrial Engineering, Purdue
University).

Atwood, M., Mora, J., & Kaplan, A. (2010). Learning to lead: Evaluating
leadership and organizational learning. Leadership & Organization
Development Journal, 31(7), 576-595.
doi:10.1108/01437731011079637

Avers, K., & Johnson, W. (2011). A review of federal aviation
administration fatigue research: Transitioning scientific results to
the aviation industry. Aviation Psychology and Applied Human
Factors, 1(2), 87–98. doi:10.1027/2192-0923/a000016

Baltar, F., & Brunet, I. (2012). Social research 2.0: Virtual snowball
sampling method using Facebook. Internet Research, 22(1), 57-74.
doi:10.1108/10662241211199960

Batteau, A. (2010). Technological peripheralization. Science,
Technology, & Human Values, 35(4) 554-574.
doi:10.1177/0162243909345834

Baxter, P., & Jack, S. (2008). Qualitative case study methodology:
Study design and implementation for novice researchers. The

Qualitative Report, 13(4), 544-559. Retrieved from
http://www.nova.edu/ssss/QR/QR13-4/baxter.pdf

Bazargan, M. & Guzhva, V. (2011). Impact of gender, age and
experience of pilots on general aviation accidents. Accident
Analysis & Prevention, 43(1), 962-970.
doi:10.1016/j.aap.2010.11.023

Bendrath, R. & Mueller, M. (2011). The end of the net as we know it?
Deep packet inspection and Internet governances. New Media &
Society, 13(7), 1142-1160. doi:10.1177/1461444811398031

Bergeon, F. & Hensley, M. (2009). Swiss cheese and the PRiMA model:
What can information technology learn from aviation accidents?
Journal of Operational Risk, 4(3), 47-58. Business Source
Complete, EBSCOhost (accessed April 8, 2013).

Berger, L., Begun, A., & Otto-Salaj, L. (2009). Participant recruitment in
intervention research: scientific integrity and cost-effective
strategies. International Journal of Social Research Methodology,
12(1), 79-92. doi:10.1080/13645570701606077

Berry, K. & Pace, J. (2011). Examining the actors and functions of an
airline operations center. Proceedings of the Human Factors and
Ergonomics Society Annual Meeting, 55(1), 1412-1416.
doi:10.1177/1071181311551294

Berry, K., Stringfellow, P., & Shappell, S. (2009). Considering trends
among industrial accidents: A preliminary meta-analysis of HFACS
causal factors across industries. In Proceedings of the Human
Factors and Ergonomics Society Annual Meeting, 53(20), 1574-
1578. SAGE Publications.

Black, R. & Chimka, J. (2011). Re-estimating and remodeling general
aviation operations. International Journal of Applied Aviation
Studies, 11(1), 47-56.

Blair, E., & O'Toole, M. (2010). Leading measures. Professional Safety,
55(8), 29-34.

Bolstad, C., Endsley, M., Costello, A., & Howell, C. (2010). Evaluation of
computer-based situation awareness training for general aviation
pilots. The International Journal of Aviation Psychology, 20(3), 269-
294.

Bowen, E., Sabin, E., & Patankar, M. (2011). Aviation maintenance
human factors in a systems context: Implications for training.
International Journal of Applied Aviation Studies, 11(1), 13-26.

Boyne, G., & Meier, K. (2009). Environmental turbulence, organizational
stability, and public service performance. Administration and
Society, 40(8), 799–824. doi: 10.1177/0095399708326333

Brooks, C. (2010). Embodied transcription: A creative method for using
voice-recognition software. The Qualitative Report, 15, 1227-1241.
Retrieved from http://www.nova.edu/ssss/QR/index.html

Cachia, M., & Millward, L. (2011). The telephone medium and semi-
structured interviews: A complementary fit. Qualitative Research in
Organizations and Management, 6(3), 265-277.
doi:10.1108/17465641111188420

Calvert, S. (2013). Gendered narratives of innovation through competition: Lessons from science and technology studies. Journal of Education for Library and Information Science, 54(1), 3-14.

Cane, S., McCarthy, R., & Halawi, L. (2010). Ready for battle? A phenomenological study of military simulation systems. The Journal of Computer Information Systems, 50(3), 33-40.

Cao, J. & Ding, S. (2012). Sensitivity analysis for safety design verification of general aviation reciprocating aircraft engine. Chinese Journal of Aeronautics, 25(1), 675-680. Retrieved from http:// www.elsevier.com/locate/cja

Carlsen, H., Dreborg, K., Godman, M., Hansson, S., Johansson, L., & Wikman-Svahn, P. (2010). Assessing socially disruptive technological change. Technology in Society, 32(1), 209-218.

Casner, S. & Schooler, J. (2014). Thoughts in flight: Automation use and pilot's task-related and task unrelated thoughts. Human Factors, 56(3), 433-442. doi:10.1177/0018720813501550

Cassens, R., Young, J., Greenan, J., & Brown, J. (2011). Elements related to teaching pilots aeronautical decision making. Collegiate Aviation Review, 29(1), 10-27.

Catino, M. & Patriotta, G. (2013). Learning from errors: Cognition, emotions and safety culture in the Italian Air Force. Organizational Studies, 0(0), 1-31. doi:10.1177/0170840612467156

Cavaoz, D., & Rutherford, M. (2011). Examining how media coverage impacts the regulatory notice and comment process. The American Review of Public Administration, 41(6) 625–638. doi:10.1177/0275074010387660.

Chandler, J. (2012). Obligatory technologies: Explaining why people feel compelled to use certain technologies. Bulletin of Science Technology & Society, 32(4), 255-264. doi:10.1177/0270467612459924

Chiappe, D., Vu, K., and Strybel, T. (2012). Situation awareness in the NextGen air traffic management system. International Journal of Human Computer Interaction, 28(1),140–151. doi:10.1080/10447318.2012.634768

Chou, H. & Zolkiewski, J. (2012). Managing resource interaction as a means to cope with technological change. Journal of Business Research, 65(1), 188-195. doi:10.1016/j.jbusres.2011.05.021

Cilesiz, S. (2011). A phenomenological approach to experiences with technology: current state, promise, and future directions for research. Educational Technology Research Development, 59(1), 487-510. doi:10.1007/s11423-010-9173-2.

Cimpian, I. (2011). Ergonomic design of aircraft cockpit. Journal of Industrial Design and Engineering Graphics, 6(1), 25-28.

Cockfield, A. (2011). Surveillance as law. Griffith Law Review, 20(4), 795-816.

Cooper, D. R., & Schindler, P. S. (2010). Business research methods. New York, NY: McGraw-Hill/Irwin.

Corallo, A., Lazoi, M., Margherita, A., & Scalvenzi, M. (2010).

Optimizing competence management processes: A case study in the aerospace industry. Business Process Management Journal, 16(2), 297-314. doi:10.1108/14637151011035615.

Craig, C. (2012). Improving flight condition situational awareness through Human centered design. Work, 41(1), 4523-4531. doi:10.3233/wor-2012-0031-4532.

Creswell, J. (2009). Research design: Qualitative, quantitative, and mixed methods approaches. (3rd ed.). Thousand Oaks, CA: SAGE Publications, Inc.

Das, A., Kumar, V., & Kumar, U. (2011). The roles of leadership competencies for implementing TQM: An empirical study in Thai manufacturing industry. International Journal of Quality and Reliability Management, 28(2) 195-219. doi:10.1108/02656711111101755

Denning, S. (2011). Reinventing management: The practices that enable continuous innovation. Strategy and Leadership, 39(3), 16-24. doi:10.1108/1087851111128775.

Dhurup, M., Mafini, C., & Malan, J. (2013). Consumer responses to salient image attributes in restaurant selection in Southern Gauteng, South Africa. Journal of Educational and Social Research, 4(3), 283-294.

Di Renzo, J. (2010). The impact of transition training on adapting to technically advanced aircraft at regional airlines: Perceptions of pilots in training and instructor pilots. Collegiate Aviation Review, 28(1), 42-52.

Dimitriadis, C. (2012). How are schools in England addressing the needs of mathematically gifted children in primary classrooms? A review of practice. Gifted Child Quarterly, 56(2), 59-76. doi:10.1177/0016986211433200

Dobson, R. (2012). Pilot salient identity influencing critical incident sensemaking in general aviation organizational safety. (Order No. 3523943, The George Washington University). ProQuest Dissertations and Theses, 165.

Dokko, G., Nigam, A., & Rosenkopf, L. (2012). Keeping steady as she goes: A negotiated order perspective on technological evolution. Organizational Studies, 33(5-6), 681-703. doi:10.1177/0170840612443624

Donahue, A., & O'Leary, R. (2011). Do shocks change organizations? The case of NASA. Journal of Public Administration Research and Theory, 22(1), 395-425. doi:10.1093/jopart/mur034

Doorn, N. (2012). Exploring responsibility rationales in research and development. Science, Technology, & Human Values, 37(3), 180-209. doi:10.1177/0162243911405344

Douglas, S. (2010). Some thoughts on the question "How do new things happen?". Technology and Culture, 51(2), 293-304. The Johns Hopkins University Press. Retrieved March 9, 2013, from Project MUSE database.

Downer, J. (2010). Trust and technology: the social foundations of

aviation regulation. The British Journal of Sociology, 61(1),83-106.
doi:10.1111/j.1468-4446.2009.01303.x

Doz, Y. (2011). Qualitative research for international business. Journal
of International Business Studies, 42(5), 582–590.
doi:10.1057/jibs.2011.18

Duncan, N. (2013). A Delphi Study of Decision Support Systems for
Aviation Safety Management Program Cost Estimation.
Northcentral University. Retrieved from
http://library.ncu.edu/ncu_diss/display_abstract.aspx?dissertation_i
d=2267

Ehrmann, S. (2011). Taking the long view: Ten recommendation about
time, money, technology, and learning. Planning for Higher
Learning, 39(2), 34-40. doi:10.1080/00091383.2010.503175

Erichsen, E., DeLorme, L., Connelley, R., Okurut-Ibore, C., McNamara,
L., Aljohani, O. (2013). Socio-technical systems approach: An
internal assessment of a blended doctoral program. The Journal of
Continuing Higher Education, 61(1), 23-34.
doi:10.1080/07377363.2013.758553

Farrell, D., & Kalil, T. (2010). United States: A strategy for innovation.
Issues in Science and Technology, 26(3), 41-50.

Feary, M., McCloy, T., Wickens, C., Kaber, D., Pritchett, A., & Sherry, L.
(2010). Bridging the gap between human: Automation interface
analysis and flight deck design guidance. Proceedings of the
Human Factors and Ergonomics Society Annual Meeting, 54(1),
36-39. doi:10.1177/154193121005400109

Federal Aviation Administration (FAA). (2010). Safety management
system (SMS) Aviation Rulemaking Committee (ARC) SMS ARC
Recommendations Final Report. Retrieved from
http://www.arsa.org/files/SMS-ARC-Final-Report-20100331.pdf

Finlay, L. (2012). Unfolding the phenomenological research process:
iterative stages of seeing afresh. Journal of Humanistic Psychology,
53(2), 172-201. doi:10.1177/0022167812453877

Flouris, T. & Yilmaz, A. (2009). Change management as a road map for
safety management system implementation in aviation operations:
Focusing on risk management and operational effectiveness.
International Journal of Civil Aviation, 1(1), 1-19.

Franza, A. & Fanjoy, R. (2012). Contributing factors in Piper 28 and
Cirrus SR20 aircraft accidents. Journal of Aviation Technology and
Engineering, 1(2), 90-96. doi:10.5703/1288284314662

Friday, M. (2011). A mixed methods study of designated pilot examiners
experiences with advanced display technology and technology
advanced aircraft. Oklahoma State University. ProQuest
Dissertations and Theses, 104.

Ganguly, A., Nilchiani, R., & Farr, J. (2010). Defining a set of metrics to
evaluate the potential disruptiveness of a technology. Engineering
Management Journal, 22(1), 34-44. Retrieved from
http://search.proquest.com.proxy1.ncu.edu/docview/734620121?ac
countid=28180

Garibay, A. & Young, J. (2013). Reducing general aviation accidents by
  utilizing airline operational strategies. Aviation Technology
  Graduate Student Publication, 1(1), 1-20. Retrieved from
  http://docs.lib.purdue.edu/atgrads

Geiselman, E., Johnson, C., & Buck, D. (2013). Flight deck automation:
  Invaluable Collaborator or Insidious Enabler? Ergonomics in
  Design, 21(1), 22-26. doi:10:1177/1064804613491268

General Aviation Joint Steering Committee Report. (2012). Loss of
  control work group: Approach and landing. Retrieved from
  http://download.aopa.org/advocacy/130327safety-committee.pdf

Giles, D., Smythe, E., & Spence, D. (2012). Exploring relationships in
  education: A phenomenological inquiry. Australian Journal of Adult
  Learning, 52(2), 214-236.

Giorgi, A. (2008). Concerning a serious misunderstanding of the
  essence of the phenomenological method in psychology. Journal of
  Phenomenological Psychology, 39(1), 33-58.
  doi:10.1163/156916208x311610

Godin, B. (2010). Innovation without the world: William F. Ogburn's
  contribution to the study of technological innovation. Minerva, 48
  (3), 277-307. doi:10.1007/s11024-010-9151-1

Goffin, K., Raja, J. Z., Claes, B., Szwejczewski, M., & Martinez, V.
  (2012). Rigor in qualitative supply chain management research.
  International Journal of Physical Distribution & Logistics
  Management, 42(8), 804-827. doi:10.1108/09600031211269767

Gordon, T., Nichter, M., & Henriksen, R. (2013). Raising black males
  from a black father's perspective: A phenomenological study. The
  Family Journal, 21(2), 154-161. doi:10.1177/1066480712466541

Gray, J. (2009). The impact of new and emerging technologies in the
  commercial aviation maintenance, repair, and overhaul industry a
  delphi study. University of La Verne). ProQuest Dissertations and
  Theses, 252.

Greenfield, B., Bridges, P., Hoy, S., Metzger, R., Obuaya, G., &
  Resutek, L. (2012). Exploring experienced clinical instructors'
  experiences in physical therapist clinical education: A
  phenomenological study. Journal of Physical Therapy Education,
  26(3), 40-47.

Halleran, M., & Wiggins, M. (2010). Changing general aviation flight
  training by implementing FAA industry training standards.
  International Journal of Applied Aviation Studies, 10(1), p. 117-130.

Hallström, J., & Gyberg, P. (2011). Technology in the rear-view mirror:
  How to better incorporate the history of technology into technology
  education. International Journal of Technology and Design
  Education, 21(1), 3-17. doi:10.1007/s10798-009-9109-5

Hancock, P., Jagacinski, R., Parasuraman, R., Wickens, C., Glenn, W.,
  Kaber, D. (2013). Human-automation interaction research: Past,
  present, and future. Ergonomics in Design: The Quarterly of Human
  Factors Applications, 21(9), 9-14. doi:10.1177/1064804613477099

Harl, R., & Roberts, P. (2011). The black experience in business

aviation: An exploratory case study. Journal of Aviation Technology
and Engineering, 1(1), 11–18. doi:10.5703/1288284314631

Harper, M., & Cole, P. (2012). Member checking: Can benefits be
gained similar to group therapy? The Qualitative Report, 17(2),
510-517.

Harris, D. (2011). Rules fragmentation in the airworthiness regulations:
A human factors perspective. Journal of Aviation Psychology and
Applied Human Factors, 1(2), 75-86. doi:10.1027/2192-
0923/a000012

Haslbeck, A., Schubert, E., Onnasch, L., Hüttig, G., Bubb, H., &
Bengler, K. (2012). Manual flying skills under the influence of
performance shaping factors. Work, 41(1), 178-183.
doi:10.3233/WOR-2012-0153-178

Hastings, S. (2010). Triangulation. In Neil J. Salkind (Ed.), Encyclopedia
of Research Design. (pp. 1538-1541). Thousand Oaks, CA: SAGE
Publications, Inc. doi:10.4135/9781412961288.n469

Hathorn, D., Machtmes, K., & Tillman, K. (2009). The lived experience
of nurses working with student nurses in the clinical environment.
The Qualitative Report, (14)2, 227-244. Retrieved from
http://www.nova.edu/sss/QR/AR14-2/hathorn.pdf

Hendrickson, S. (2009). The wrong Wright stuff: Mapping human error
in aviation. (Doctoral dissertation, University of New Mexico, 2009).
Dissertation Abstracts International, 70/06.

Hughes, J., Rice, S., Trafimow, D., & Clayton, K. (2009). The automated
cockpit: A comparison of attitudes towards human and automated
pilots. Transportation Research Part F: Traffic Psychology and
Behavior, 12(5), 428-439. doi:10.1016/j.trf.2009.08.004

Hycner, R. (1985). Some guidelines for the phenomenological analysis
of interview data. Human Studies, 8(1), 279-303.

Jackson, J. (2010). Apparently safe: How aviation regulation
characterizes risk. (University of Washington). ProQuest
Dissertations and Theses, 242.

Jackson, S. & Philip, G. (2010). A techno-cultural emergence
perspective on the management of techno-change. International
Journal of Information Management, 30(1), 445–456.
doi:10.1016/j.ijinfomgt.2010.01.008

Jalonen, H. (2012). The uncertainty of innovation: A systematic review
of the literature. Journal of Management Research, 4(1), 1-48.
doi:10.5296/jmr.v4i1.1039

Kapoulas, A., & Mitic, M. (2012). Understanding challenges of
qualitative research: Rhetorical issues and reality traps. Qualitative
Market Research, 15(4), 354-368.
doi:10.1108/13522751211257051

Kearns, S. and Sutton, J. (2013). Hangar talk survey: Using stories as a
naturalistic method of informing threat and error management
training. The Journal of the Human Factors and Ergonomics
Society, 55(2), 267-277. doi:10.1177/0018720812452127

Kikooma, J. F. (2010). Using qualitative data analysis software in a

social constructionist study of entrepreneurship. Qualitative
Research Journal, 10(1), 40-51.

Kim, Y. (2010). The pilot study in qualitative inquiry: Identifying issues
and learning lessons for culturally competent research. Qualitative
Social Work, 10(2), 190-206. doi.10.1177/1473325010362001

King, G. (2011). General aviation training for automation surprise.
International Journal of Professional Aviation Training & Testing
Research, 5(1), 46-51. Retrieved from
http://ojs.library.okstate.edu/osu/index.php/IJPATTR/article/view/43
0

Knecht, W. (2013). The killing zone revisited: Serial nonlinearities
predict general aviation accident rates from pilot total flight hours.
Accident Analysis and Prevention, 60(1), 50-56. doi:
10.1016/j.aap.2013.08.012

Kvale, S. and Brinkman, S. (2009). Interviews: Learning the craft of
qualitative research interviewing. (2nd ed.). Thousand Oaks, CA:
SAGE Publications, Inc.

Lakshmi, S. & Mohideen, M. (2013). Issues in reliability and validity of
research. International Journal of Management Research and
Reviews, 3(4), 2752-2758.

Landeweerd, L., Osseweijer, P., & Kinderlerer, J. (2009). Distributing
responsibility in the debate on sustainable biofuels. Science and
Engineering Ethics, 15(4), 531-543. doi: 10.1007/s11948-009-9154-
1

Lane, C. & Lyle, H. (2011). Obstacles and supports related to the use of
educational technologies: The role of technological expertise,
gender, and age. Journal of Computing in Higher Education, 23(1),
38-59. doi:10.1007/s12528-010-9034-3

Layton, G. (2012). Aviation safety: Comparing national and regional
governmental regulatory commercial oversight affiliations.
International Journal of Business and Social Science, 3(3), 83-95.

Lee Y., Jeon, D., & Choi, Y. (2012). Air traffic controllers' situation
awareness and workload under dynamic air traffic situations.
Transportation Journal, 51(3), 338-352. doi:10.1353/tnp.2012.0021.

Leedy, P. & Ormrod, J. (2005). Practical research: Planning and design
(8th ed). Upper Saddle River, NJ: Pearson Education.

Levasseur, R. (2011). Dissertation research: An integrative approach.
St. Augustine, FL: MindFire Press.

Leveson, N. (2011). Applying systems thinking to analyze and learn
from events. Safety Science 49, 55-64.
doi:10.1016/j.ssci.2009.12.021

Lewis, C. (2014). Examining principle core competencies for aviation-
safety professionals. Northcentral University. ProQuest
Dissertations and Theses, 146.

Liang, T. (2010). Innovative sustainability and highly intelligent human
organizations: The new management and leadership perspective.
International Journal of Complexity in Leadership and
Management, 1(1), 83-109.

Lindo, R., Deaton, J., Cain, J., & Lang, C. (2012). Methods of instrument
training and effects on pilots' performance with different types of
flight instrument displays. Journal of Aviation Psychology and
Applied Human Factors, 2(2), 62-71. doi:10.10272192-
0923/a000028.

Lu, C., Young, J., Schreckengast, S., & Chen, H. (2011). Safety culture:
The perception of Taiwan's aviation leaders. International Journal
of Applied Aviation Studies, 11(1), 27-46.

Madsen, P. (2013). Perils and profits: A re-examination of the link
between profitability and safety in U.S. Aviation. Journal of
Management, 39(3), 763-791. doi:10.1177/0149206310396374

Mangioni, V., & McKerchar, M. (2013). Strengthening the validity and
reliability of the focus group as a method in tax research. EJournal
of Tax Research, 11(2), 176-190.

Manzey, D., Reichebach, J., & Onnasch, L. (2012). Human performance
consequences of automated decision aids: The impact of degree of
automation and system experience. Journal of Cognitive
Engineering and Decision Making, 6(1), 57-87.
doi:10.1177/1555343411433844

Marshall, B., Cardon, P., Poddar, A., & Fontenot, R. (2013). Does
sample size matter in qualitative research?: A review of qualitative
interviews in is research. The Journal of Computer Information
Systems, 54(1), 11-22.

Mavin, T., & Murray, P. (2010). The development of airline pilot skills
through simulated practice. Learning Through Practice,
Professional and Practice-based Learning, 1(1), 268-286.
doi:10.1007/978-90-481-3939-2_15

McCloy, T., Harper-Sciarini, M., Durso, F., Jentsch, F., Kanki, B., &
Rogers, W. (2011). Framing the Design, Evaluation, and
Certification Process for NextGen Technologies and Procedures
Automation Roles and Responsibilities. In Proceedings of the
Human Factors and Ergonomics Society Annual Meeting, 55(1),
120-122. SAGE Publications.

McCracken, C. (2011). Flight Training Success in Technologically
Advanced Aircraft. Aviation Technology Graduate Student
Publications. Paper 7.

McNeely, S. (2012). Examining the relationship between organizational
safety and culture and safety management system implementation
in aviation. Northcentral University. ProQuest Dissertations and
Theses, 146.

Mikel-Brumfield, N. (2009). Who are we serving? A look at the customer
of the federal government. The Journal of Quality and Participation,
32(2), 9-12.

Misa, T. (2009). Findings follow framings: Navigating the empirical turn.
Synthese, 168(3), 357-375. doi:10.1007/s11229-008-9447-y

Mitchell, J., Vermuelen, L., & Naidoo, P. (2009). Flying glass: A
qualitative analysis of pilot perceptions of automated flight-decks
after 20 years. International Journal of Applied Aviation Studies,

9(1), 13-28.

Moiser, K., Fischer, U., Morrow, D., Feigh, K., Durso, F., Sullivan, K., &
Pop, V. (2013). Automation, task, and context features: Impacts on
pilots' judgments of human–automation interaction. Journal of
Cognitive Engineering and Decision Making, 7(4), 377-399.
doi:10.1177/1555343413487178

Mosier, K. L., Fischer, U., Cunningham, K., Munc, A., Reich, K., Tomko,
L., & Orasanu, J. (2012). Aviation Decision Making Issues and
Outcomes: Evidence from ASRS and NTSB Reports. In
Proceedings of the Human Factors and Ergonomics Society Annual
Meeting, 56(1), 1794-1798. Sage Publications.

Moustakas, C. (1994). Phenomenological research methods. Thousand
Oaks, CA: SAGE Publications Inc.

National Transportation Safety Board (NTSB). (2010). Introduction of
glass cockpits into avionics into light aircraft. Safety Study
(NTSB/SS-01/10, PB2010917001). Retrieved from
http://www.ntsb.gov/publictn/2010/SS1001.pdf

Naveh, E., Katz-Navon, T., & Stern, Z. (2011). The effect of safety
management systems on continuous improvement of patient safety:
The moderating role of safety climate and autonomy. The Quality
Management Journal, 18(1), 54-67.

Nilsson, S. (2011). Relationship between recent flight experience and
pilot error general aviation accidents. (Doctoral dissertation,
Northcentral University, 2011). Dissertation Abstracts International,
72/09.

Norlyk, A., & Harder, I. (2010). What makes a phenomenology study
phenomenological? An analysis of peer-reviewed empirical nursing
studies. Qualitative Health Research, 20(3), 420-431.
doi:10.1177/1049732309357435

Novicevic, M., Hayek, M., Buckley, M., & Humphreys, J. (2009).
Chandler and technological determinism in the histories of
management. The Journal of Applied Management and
Entrepreneurship, 14(4), 3-13.

O'Toole, M., & Nalbone, D. (2011). Safety perception surveys: What to
ask, how to analyze. Professional Safety, 56(6), 58-62.

O'Connor, P., O'Dea, A., Kennedy, Q., & Buttrey, S. (2011). Measuring
safety climate in the aviation industry: A review and
recommendations for the future. Safety Science, 49, 128-138.

Ofonedu, M., Percy, W., Harris-Britt, A., & Belcher, H. (2013).
Depression in inner city African American youth: A
phenomenological study. Journal of Children Family Studies, 22(1),
96-106. doi:10.1007/s10826-012-9583-3

Oliver, M. (2011). Technological determinism in educational technology
research: some alternative ways of thinking about the relationship
between learning and technology. Journal of Computer Assisted
Learning, 1(27), 373-384. doi:10.1111/j.1365-2729.2011.00406.x

Oluwatayo, J. (2012). Validity and reliability issues in educational
research. Journal of Educational and Social Research, 2(2), 391-

400.

Oster, C., Strong, J. & Zorn, K. (2013). Analyzing aviation safety: Problems, challenges, opportunities. Research in Transportation Economics, 43(1), 148-164. Retrieved from http://www.elsevier.com/locate/retrec

Oztekin, A., & Luxhøj, J. (2010). An inductive reasoning approach for building system safety risk models of aviation accidents. Journal of Risk Research, 13(4), 479-499. doi:10.1080/13669870903484344

Paletz, S., Bearman, C., Orasanu, J., & Holbrook, J. (2009). Socializing the human factors analysis and classification system: Incorporating social psychological phenomena into a human factors error classification system. The Journal of the Human Factors and Ergonomics Society, 51(4), 435-445. doi:10.1177/0018720809343588

Parasuraman, R. & Manzey, D. (2010). Complacency and bias in human use of automation: An attentional integration. Journal of the Human Factors and Ergonomics Society, 52, (1), 381-410. doi:10.1177/0018720810376055

Parchoma, G. (2009). Toward developing teaching, learning, and technology perspectives-in-practice for networked learning. In T. Bastiaens et al. (Eds.), Proceedings of World Conference on E-Learning in Corporate, Government, Healthcare, and Higher Education 2009 (p. 3061-3068). Chesapeake, VA: AACE.

Parylo, O., Zepeda, S., & Bengston, E. (2012). Principals' experiences of being evaluated: A phenomenological study. Education Assessment Evaluation Accountability, 24(1), 215-238. doi:10.1007/s11092-012-9150-x

Patton, M. (2002). Qualitative research & evaluation methods (3rd ed.). Thousand Oaks, CA: Sage.

Pettersen, K., McDonald, N., & Engen, O. (2010). Rethinking the role of social theory in socio-technical analysis: A critical realist approach to aircraft maintenance. Cognition, Technology and Work, 12(3), 181-191. doi:10.1007/s10111-009-0133-8

Plankey-videla, N. (2012). Informed consent as process: Problematizing informed consent in organizational ethnographies. Qualitative Sociology, 35(1), 1-21. doi:10.1007/s11133-011-9212-2

Pourdehnad, J. & Smith, P. (2012). Sustainability, organizational learning, and lessons learned from aviation. The Learning Organization, 19(1), 77-86. doi:10.1108/09696471211190374

Pritchett, A. (2009). Aviation automation: General perspectives and specific guidance for the design of modes and alerts. Reviews of Human Factors and Ergonomics, 5(1), 82-113. doi:10.1518/155723409X448026

Randles, C. (2012). Phenomenology: A review of literature. National Association for Music Education, 3(20), 11-21. doi:10.1177/8755123312436988

Reynolds, R., Blickensderfer, E., Martin, A., Rossignon, K., & Maleski, V. (2010). Human factors training in aviation maintenance: Impact

on incident rates. Human Factors: The Journal of the Human
Factors and Ergonomics Society, 54(19), 1518-1520.

Rice, S. (2009). Examining single and multiple-process theories of trust
in automation. The Journal of General Psychology, 136(3), 303–
319. doi:10.3200/GENP.136.3.303-322

Ring, N., Ruth, J., & Ritchie, K. (2011). Methods of synthesizing
qualitative research studies for health technology assessment.
International Journal of Technology Assessment in Health Care,
27(4), 384–390. doi:10 1017/S0266462311000389

Robertson, C. (2010). Determining appropriate levels of automation.
Retrieved from
http://www.faa.gov/training_testing/training/fits/.../Det_App_Lvl_Atm
.pdf

Ropp, T. (2009). Implementing safety management systems for aviation
into an aviation technology curriculum. The Technology Interface
Journal, 10(2). Retrieved from http://technologyinterface.nmsu.edu

Rutherford-Hemming, T., Vlasses, F., & Rogers, J. (2012). Practice
makes perfect: Tips for successful institutional review board
submissions. The Journal of Continuing Education in Nursing,
43(5), 203-8; quiz 209-10. doi:10.3928/00220124-20111101-03

Sadasivan, S. & Gramopadhye, A. (2009). Technology to support
inspection training in the general aviation industry: Specification
and design. International Journal of Industrial Ergonomics, 39(1),
608-620. doi:10.1016/j.ergon.2008.09.002

Sanderson, H., & Lea, J. (2012). Implementation of the clinical
facilitation model within an Australian rural setting: The role of the
clinical facilitator. Nurse Education in Practice, 12(6), 333-9.
doi:10.1016/j.nepr.2012.04.001

Sawyer, M., Berry, K., & Blanding R. (2011). Assessing the human
contribution to risk in NextGen. Proceedings of the 55th Annual
Meeting of the Human Factors and Ergonomics Society. San
Francisco, CA. doi:10.1177/ 1071181311551012

Schram, T. (2006). Conceptualizing and proposing qualitative research
(2nd ed.). Upper Saddle River, NJ: Pearson Education.

Seidman, I. (2006). Interviewing as qualitative research (3rd ed.). New
York, NY: Teachers College Press.

Selwyn, N. (2012). Making sense of young people, education and digital
technology: The role of sociological theory. Oxford Review of
Education, 38(1), 81-96. doi:10.1080/03054985.2011.577949

Shao, B., Guindani, M., & Boyd, D. (2014). Fatal accident rates for
instrument rated private pilots. Aviation, Space, and Environment
Medicine, 85(6), 631-637.

Sharma, V., Coit, D., Oztekin, A., & Luxor, J. (2009). A decision analytic
approachfor technology portfolio prioritization: aviation safety
applications. Journal of Risk Research, 12(6), 843-864.
doi:10.1080/13669870902970236.

Shetty, K. I. (2012). Current and historical trends in general aviation in
the United States (Doctoral dissertation, Massachusetts Institute of

Technology Cambridge, MA 02139 USA).

Smale, M. A. (2010). Demystifying the IRB: Human subjects' research in academic libraries. Portal: Libraries and the Academy, 10(3), 309-321.

Smith, C. (2008). Glass cockpit transition training in collegiate aviation: Analog to digital. Retrieved from: http://etd.ohiolink.edu/sendpdf.cgi/Smith%20Catherine.pdf?bgsu12 25479328

Smith, T. (2011). Safety management: A personal development strategy. Professional Safety, 56(3), 58-68.

Sobieralski, J. (2013). The cost of general aviation accidents in the United States. Transportation Research Part A: Policy and Practice, 47, (1), 19-27. doi:10.1016/j.tra.2012.10.018.

Spiri, W., & MacPhee, M. (2013). The meaning of evidence-based management to brazilian senior nurse leaders. Journal of Nursing Scholarship, 45(3), 265-72.

Stav, W. (2012). Developing and implementing driving rehabilitation programs: A phenomenological approach. The American Journal of Occupational Therapy, 66(1), e11-9.

Stolzer, A. (2011). The rationale for Embry Riddle Aeronautical University's Ph.D. in aviation: Brief explanation. The Journal of Aviation Aerospace Education and Research, 20(2), 9-11. Retrieved from http://commons.erau.edu/jaaer/vol20/iss2/5

Strauch, B. (2010). Can cultural differences lead to accidents? Team cultural differences and socio-technical system operations. Journal of Human Factors and Ergonomics Society, 52(2), 246-263. doi:10.1177/0018720810362238

Sullivan, B. (2010). Competition and beyond: Problems and attention allocation in the organization rulemaking process. Organization Science, 21(2), 432-450. doi:10,1287/orsc.1090.0436.

Suri, H. (2011). Purposeful sampling in qualitative research synthesis. Qualitative Research Journal, 11(2), 63-75.

Talbert, P. Y. (2012). Strategies to increase enrollment, retention, and graduation rates. Journal of Developmental Education, 36(1), 22-24,26-29,31,33,36.

Taylor, G., Reinerman-Jones, L., Szalma, J., Mouloua, M., & Hancock, P. (2013). What to automate: Addressing the multidimensionality of cognitive resources through system design. Journal of Cognitive Engineering and Decision Making, (1), 1-19. doi:10.1177/1555343413495396

Tirgari, V. (2012). Information technology policies and procedures against unstructured data: A phenomenological study of information technology professionals. Academy of Information and Management Sciences Journal, 15(2), 87-106.

Townsend, A., Cox, S. M., & Li, L. C. (2010). Qualitative research ethics: Enhancing evidence-based practice in physical therapy. Physical Therapy, 90(4), 615-28.

Tufford, L. & Newman, P. (2010). Bracketing in qualitative research.

Qualitative Social Work, 11(1), 80-96.
doi:10.1177/1473325010368316.
Twigg, D. (2011). Handle with care: Researching the lived experiences
of young children in early childhood settings. International Journal
of Arts & Sciences, 4(11), 169-178.
United States Department of Transportation (USDOT). (2011, April).
The Future Aviation Advisory Committee: FAAC Final Report.
Retrieved from http://www.dot.gov/highlights/future-aviation-
advisory-committee/faac-final-report.
United States Department of Transportation (USDOT). (2012, April) The
state of aviation safety and FAA's oversight of the national airspace
system. (Publication No. CC-2012-018). Before the Committee on
Transportation and Infrastructure Subcommittee on Aviation.
Statement by Jeffrey B. Guzzetti, Assistant Inspector General for
Aviation and Special Program U.S. Department of Transportation.
United States Government Accountability Office (GAO). (2011,
November). Initial pilot training: Better management controls are
needed to improve FAA oversight. (Publication No. GAO-12-117).
Retrieved from: http://www.gpoaccess.gov/gaoreports/index.html
United States Government Accountability Office (GAO). (2012,
October). General aviation safety: Additional FAA efforts could help
identify and reduce safety risks. (Publication No. GAO-13-36).
Washington, DC: Gerald Dillingham, Brandon Haller, Pamela
Vines, Jessica Wintfeld, Russ Burnett, Bert Japikse, Delwen Jones,
Josh Ormond, and Jeff Tessin. Retrieved from:
http://www.gpoaccess.gov/gaoreports/index.html
Vanderburg, W. (2012). Moving beyond technological determinism and
autonomy to face our responsibilities. Education Technology, 52(1),
25-31.
Vasconcelos, A. F. (2010). The effects of prayer on organizational life: A
phenomenological study. Journal of Management and
Organization, 16(3), 369-381.
Vincent, M., Blickensderfer, E., Thomas, R., Smith, M., & Lanicci, J.
(2013, September). In-Cockpit NEXRAD Products Training General
Aviation Pilots. In Proceedings of the Human Factors and
Ergonomics Society Annual Meeting, 57(1), 81-85. SAGE
Publications.
Wachter, J. K. (2011). Ethics: The absurd yet preferred approach to
safety management. Professional Safety, 56(6), 50-57.
Way, A., & Marques, J. (2013). Management of gender roles: Marketing
the androgynous leadership style in the classroom and the general
workplace. Organization Development Journal, 31(2), 82-94.
Webster, M. (2013). Philosophy of technology assumptions in
educational technology leadership: Questioning technological
determinism. (Order No. 3569909, Northcentral University).
ProQuest Dissertations and Theses, 325.
Welch, C., Piekkari, R., Plakoyiannaki, E., & Paavilainen-mäntymäki, E.
(2011). Theorising from case studies: Towards a pluralist future for

international business research. Journal of International Business Studies, 42(5), 740-762. doi:10.1057/jibs.2010.55

Whitehurst, G., & Rantz, W. (2013). The digital to analog risk: Should we teach new dogs old tricks? Journal of Aviation Aerospace Education and Research, 21(3), 17-22.

Wright, D. (2012). Redesigning informed consent tools for specific research. Technical Communication Quarterly, 21(2), 145-167.

Yang, G. (2009). The internet as cultural form: Technology and the human condition in China. Knowledge, Technology, and Politics, 22(1), 109–115. doi:10.1007/s12130-009-9074-z

Yang, J., Kennedy, Q., Sullivan, J., & Fricker, R. (2013). Pilot performance: How scan patterns and navigational assessments vary by flight expertise. Aviation, Space, and Environmental Medicine, 84(2), 116-124.

Yeh, M., Swider, C., Abbott. K., Donovan, C., Neiderman, E., & Piccione, D. (2012). Human factors at the Federal Aviation Administration (FAA): From research to reality. Proceedings of the Human Factors and Ergonomics Society Annual Meeting, 56(1), 56-60. doi:10.1177/1071181312561032

Zabloski, J. & Milacci, F. (2012). Gifted dropouts: Phenomenological case studies of rural gifted students. Journal of Ethnographic and Qualitative Research, 6(1), 175-190.

Zikmund, W. (2003). Business research methods (7th ed.). Mason, OH: Thompson/South-Western.

Zucker, D. M. (2001, June). Using case study methodology in nursing research. The Qualitative Report [On-line serial], 6(2). Available: http://www.nova.edu/ssss/QR/QR6-2/zucker.html

APPENDICES

APPENDIX A

Data Analysis Procedures

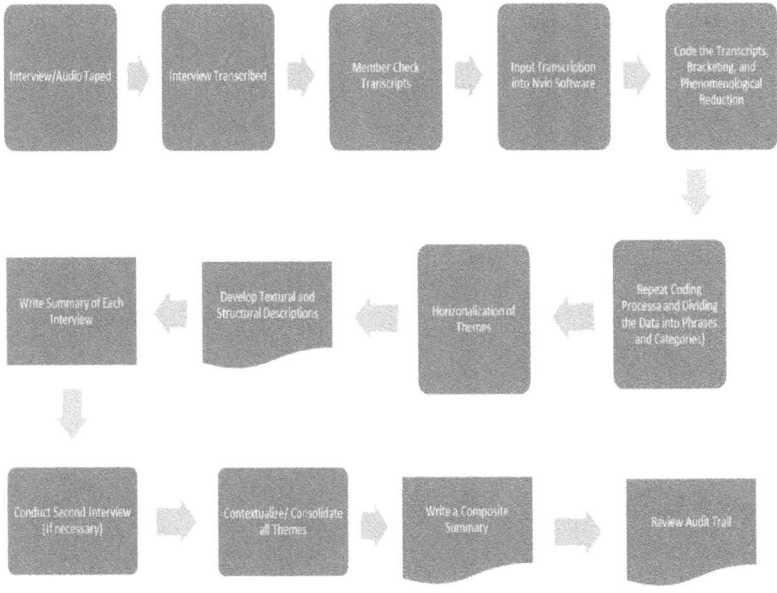

APPENDIX B

Interview Question Guide

Research ID Code: _____

Date: _____

Phone: _____

Demographic Questions:

1. What is your age?

2. How long have you been licensed as a general aviation pilot?

3. How many flight hours have you earned? How many flight hours in TAA?

4. What is your affiliation with general aviation? (i.e. pilot, CFI, designated pilot examiner)

5. What is the highest flight rating you have earned?

Opinions and Value Questions

6. How would you describe your flight training to prepare you to fly TAA?

7. Describe your experiences of how TAA impact your approach to flying?

8. What changes can the government and industry make to better prepare general aviation pilots to transition between conventional aircraft and TAA?

9. How do you view the regulatory approach in general aviation regarding TAA?

10. What is the best approach to integrate technologies into general aviation?

11. What is your view on of FITS?

12. How do you use Single Pilot Resource Management in general aviation?

13. How do you use aeronautical decision-making training in your general aviation flying?

Feelings Questions

14. How do you feel about the integration of new

    technologies in the general aviation?

15. How do you feel about the safety implications

    caused by integrating TAA in general aviation?

Appendix C

Informed Consent Form

Exploring Pilots' Experiences of Integrating Technologically
Advanced Aircraft Within General Aviation

What is the study about?

You are invited to participate in a research study being
conducted for a dissertation at Northcentral University in
Prescott, Arizona. The study is interested in your thoughts
and opinions about glass cockpit aircraft in general
aviation. You were selected because you responded to a
flyer about the study. There is no deception in this study.

What will be asked of me?

You will be asked to participate in a 60-minute interview
that will be videotaped with your approval to answer some
open-ended questions.

Who is involved?

The following people are involved in this research project
and may be contacted at any time: Calvin Nobles and Dr.
Robert George.

Are there any risks?

Although there are no known risks in this study, some of
the questions might be personally sensitive since some of

the questions ask about the meaning of life. This can be distressing to some people. However, you may stop the study at any time. You can also choose <u>not</u> to answer any question that you feel uncomfortable in answering.

What are some benefits?

There are no direct benefits to you of participating in this research. No incentives are offered. The results will have scientific interest that may eventually have benefits for general aviation.

Is the study anonymity/ confidential?

The data collected in this study are confidential. Your name or personal information is not linked to data. Only the researchers in this study will see the data.

Can I stop participating the study?

You have the right to withdraw from the study at any time without penalty. You can skip any questions on any questionnaires if you do not want to answer them.

Research Participant's Rights or Complaints.

Direct your questions about research participant's rights, complaints, or any problems to the researchers listed on the consent form. It is your right to contact Northcentral University's Institutional Review Board at irb@ncu.edu or 1-888-327-2877 ex 8014 to discuss any issue or concern. We would be happy to answer any question that may arise about the study. Please direct your questions or comments to: Calvin Nobles, PhD Candidate ███████████ [2]cn897

2@gmail.com or Dr. George, Dissertation Committee Chair

████████████████████████████

Signatures

I have read the above description for the Exploring Pilots' Experiences of Integrating Technologically Advanced Aircraft Within General Aviation Study. I understand what the study is about and what is being asked of me. My signature indicates that I agree to participate in the study.

Participant's Name :   _____

Researcher's Name: _____

Participant's Signature: _____

Researcher's Signature: _____

Date: _____

---

[2] *Direct POC information redacted for privacy of individual.*

Appendix D

Research Study and Participant Notification

October xx, 2014

From: Calvin Nobles
To: Prospective Pilot Study Research Participants
Re: Request Participation in a Pilot Study

1. I request your participation in a pilot study for dissertation research. The working title of the research study is "Exploring Pilots' Experiences of Integrating Technologically Advanced Aircraft Within General Aviation: A Case Study."

2. This pilot and research study is being conducted for a dissertation at Northcentral University in Prescott, Arizona. The purpose of this study is to explore the impacts of technological determinism within the general aviation sector. Technological deterministic thinking directly affects the social and cultural norms of an organization. Currently, the impact of technological determinism on general aviation is unknown. I am interested in opinions and reflections about your experiences with technologically advanced aircraft in general aviation.

3. Particularly, I am interested in interviewing pilots in Georgia, South Carolina, or Florida with flying experience in technologically advanced aircraft. I am asking you to participate in a face-to-face interview that will last 60 minutes. If you agree to an interview, you will be briefed on the purpose of the study, the informed consent process, the rights of a research participant, and the process to request a debrief on the study results. If you are interested, send me an e-mail at cn8972@gmail.com or call ██████████████ The data collected will be confidential and your identity will remain anonymous.

4. You have my authorization to disseminate this notification further and to post to your social media sites.

5. This study does not offer any incentives or benefits but your participation might contribute to improving general aviation safety.

6. If you have any questions, please feel free to contact me at cn8972@gmail.com or ██████████████

Sincerely,

Calvin Nobles, PhD Candidate

Appendix E

Invitation Letter to the Panel of Experts

July 12, 2014

From: Calvin Nobles
To: Prospective Expert Panel Member
Subj: Invitation to Serve as a Panel Expert for a Research Study

1. My name is Calvin Nobles and I am currently pursuing a PhD in Business Administration with a specialization in Management and Engineering Technology at Northcentral University, Prescott Arizona. The focus of my research is to achieve a comprehensive understanding on the impacts of integrating technologically advanced aircraft in general aviation and determining how to use glass cockpits to improve aviation safety and pilots' decision-making skills. The research method is a qualitative single descriptive case study. To ensure that the interview questions are valid for collecting the desired data, I would like to engage a panel of three experts in the fields of Aviation Safety, Aviation Regulation, and General Aviation.

2. Based on your background and professional expertise your contribution to this study is invaluable. I respectfully request you to serve on this panel of experts. You will be asked to review the proposed interview question guide and evaluate how effective is each interview question. Your constructive criticism should provide assistance in

developing conclusions relating to the Interview Question Guide. Please use the attached WORD document and the Track Change features so I can identify your proposed changes. I plan to begin the interviews on XXXXX, XX, 2014 and would greatly appreciate if you could submit your response by no later than XXXXX, XX, 2014. Please save the file with the proposed changes and send it via e-mail as an attachment to cn8972@gmail.com.

3. Thank you in advance for taking the time to review this study and making worthwhile recommendations. Your contribution to the field of business and management studies is greatly appreciated and will be recognized in my dissertation manuscript. This research is being done under the purview of my Dissertation Chair, Dr. Robert George, C. Fore, PhD. If you need to contact him for any reason, his email address is ███████████████.

Very Respectfully,

School of Business and Technology Management
Northcentral University, Prescott Valley, Arizona

INDEX

accidents14, 17, 19, 20, 22,
24, 25, 26, 28, 32, 38, 45,
47, 50, 56, 59, 65, 67, 68,
73, 74, 75, 78, 84, 88, 90,
92, 93, 95, 96, 100, 102,
104, 107, 112, 114, 115,
119, 120, 123, 130, 137,
141, 145, 147, 152, 153,
157, 161, 166, 178, 180,
181, 184, 191, 214, 215,
243, 253, 259, 261, 263,
271, 273, 274, 276, 283,
285, 288, 298, 300, 301,
302, 303, 304, 306, 308,
309, 312, 315, 318, 321,
325, 335, 338, 339, 340,
341, 343, 344, 350, 352,
353, 354, 355, 357, 358,
361, 364, 367, 368, 372,
373, 375
adaptable automation 124
aerial surveying 17
aeronautical decision-
making13, 59, 104, 105,
235, 268, 281, 295, 308,
311, 316, 317, 318, 322,
324, 325, 346, 360
aeronautical examination.17
agricultural 43
air traffic control78, 83, 106,
111, 121, 164, 349
air traffic controllers42, 61,
64, 117, 121, 146, 341
air transport industry 90

aircraft13, 17, 18, 22, 23,
24, 25, 27, 31, 45, 57, 58,
60, 61, 62, 63, 64, 66, 68,
73, 78, 80, 81, 83, 89, 92,
94, 95, 96, 103, 104, 108,
109, 110, 114, 118, 123,
126, 127, 128, 131, 133,
135, 136, 139, 140, 141,
142, 156, 163, 178, 183,
199, 206, 230, 238, 241,
245, 246, 247, 252, 253,
256, 258, 261, 262, 263,
266, 268, 271, 275, 276,
277, 280, 284, 285, 287,
288, 290, 296, 297, 298,
299, 301, 303, 304, 305,
306, 307, 308, 310, 311,
317, 319, 320, 322, 324,
333, 334, 336, 337, 344,
349, 353, 354, 360, 365,
366, 367, 372, 373, 384,
387, 391, 392, 393
airline transport carrier.....83
airspace architecture 106
analog gauges 61, 126
analysis-based
assessments 122
anthropology....................36
application of science 38
audit trail 55, 197, 220
automated technologies56,
267, 340
automation13, 19, 22, 24,
28, 29, 44, 49, 50, 53, 57,

58, 60, 61, 67, 68, 71, 72,
73, 79, 81, 82, 84, 85, 88,
89, 90, 91, 92, 93, 94, 96,
110, 118, 121, 123, 126,
127, 128, 129, 135, 138,
139, 141, 142, 145, 150,
153, 175, 178, 179, 181,
182, 186, 188, 198, 200,
204, 205, 214, 233, 235,
251, 253, 258, 263, 265,
268, 273, 277, 278, 279,
281, 284, 285, 289, 294,
300, 302, 303, 305, 309,
313, 314, 319, 322, 324,
325, 326, 328, 329, 331,
335, 337, 338, 339, 341,
344, 347, 350, 353, 355,
356, 357, 359, 368, 370,
371, 372, 373, 374

Automation management 60
automation surprise89, 90,
96, 128, 140, 313, 322,
325, 350, 370
autopilot 20, 76
autopilot22, 63, 91, 92, 94,
263, 265, 271, 286, 289,
302, 305
aviation12, 13, 14, 17, 18,
20, 21, 23, 24, 25, 27, 29,
35, 37, 38, 39, 41, 44, 45,
47, 49, 50, 51, 52, 53, 54,
56, 58, 60, 62, 64, 65, 67,
68, 71, 72, 73, 74, 75, 78,
79, 81, 83, 85, 86, 88, 89,
91, 92, 93, 94, 95, 96, 99,
100, 102, 104, 105, 106,
107, 108, 109, 111, 112,
113, 114, 115, 116, 118,
119, 120, 122, 123, 125,
127, 128, 129, 130, 131,
132, 134, 137, 138, 139,
140, 141, 142, 143, 145,
146, 147, 148, 149, 150,
151, 153, 156, 157, 159,

160, 162, 163, 165, 166,
167, 168, 170, 172, 173,
175, 177, 179, 181, 182,
183, 185, 186, 187, 188,
189, 190, 191, 192, 193,
197, 198, 199, 200, 201,
202, 203, 205, 206, 207,
210, 211, 213, 215, 216,
219, 222, 227, 228, 229,
230, 231, 233, 235, 237,
238, 239, 240, 241, 242,
243, 249, 251, 252, 253,
254, 256, 258, 259, 260,
261, 262, 263, 265, 267,
270, 273, 274, 276, 277,
279, 280, 281, 282, 284,
285, 286, 288, 289, 290,
293, 294, 295, 297, 298,
299, 300, 301, 302, 303,
304, 305, 306, 308, 309,
310, 312, 313, 314, 315,
316, 318, 320, 321, 323,
324, 325, 326, 327, 328,
329, 330, 333, 335, 336,
338, 339, 340, 341, 343,
344, 345, 346, 347, 348,
349, 351, 352, 353, 354,
355, 356, 357, 359, 360,
363, 364, 365, 366, 367,
368, 369, 370, 371, 372,
373, 374, 375, 376, 383,
384, 385, 387, 388, 391,
392, 393, 431, 432
aviation accidents21, 22,
27, 46, 57, 58, 60, 73, 74,
76, 79, 95, 100, 107, 112,
120, 123, 147, 152, 157,
161, 179, 214, 215, 259,
263, 285, 302, 311, 342,
343, 354
aviation industry14, 21, 26,
69, 75, 78, 80, 81, 96,
100, 104, 111, 114, 115,

118, 130, 134, 141, 154,
159, 160, 161, 167, 170,
175, 177, 185, 296, 300,
308, 336, 343, 347, 348,
356, 358, 363
aviation infrastructure 39, 43
aviation leaders48, 117, 147
aviation maintenance
personnel...........101, 120
aviation organizations59,
69, 75, 107, 112, 133,
138, 142, 143, 154, 169,
185, 200
aviation pilots13, 23, 24, 28,
29, 49, 51, 52, 57, 62, 67,
73, 74, 75, 82, 83, 87, 90,
91, 95, 103, 104, 108,
109, 112, 128, 135, 140,
141, 147, 159, 175, 177,
180, 186, 192, 197, 198,
200, 202, 203, 204, 206,
210, 211, 213, 227, 229,
230, 233, 235, 239, 249,
262, 264, 271, 281, 282,
295, 297, 306, 310, 311,
312, 313, 316, 321, 322,
330, 333, 337, 339, 340,
341, 346, 348, 350, 353,
354, 356, 360, 361
aviation safety12, 14, 19,
20, 25, 26, 28, 29, 45, 47,
71, 78, 88, 111, 115, 141,
152, 179, 215, 235, 252,
257, 299, 315, 326, 327,
334, 335, 337, 339, 342,
343, 350, 353, 356, 357,
358, 359, 360, 361
Aviation Safety Officer ...115
Aviation Safety Program60,
113
aviation sector17, 21, 24,
25, 27, 57, 68, 69, 72, 73,
76, 81, 84, 88, 99, 102,

104, 110, 112, 117, 153,
154, 169, 170, 171, 173,
175, 178, 198, 329
aviation socio-technical
systems 43
aviation specialists...42, 159
aviation stakeholders167,
175
aviation studies................99
aviation weather 130
avionics18, 61, 91, 92, 111,
118, 128, 139, 266, 272,
284, 289, 337, 344, 372
business practices36, 117,
154, 168, 172
case study13, 29, 49, 50,
54, 58, 67, 71, 179, 182,
183, 185, 186, 187, 188,
190, 191, 193, 196, 197,
198, 203, 216, 219, 225,
227, 232, 233, 235, 237,
242, 294, 326, 329, 331,
332, 349, 356, 359, 363,
366, 369, 377, 393
cathode ray tube
CRT 126
certified flight instructors29,
54, 110, 163, 169, 186,
199, 201, 237, 241, 247,
249, 251, 260, 293, 295,
296, 309, 316, 327, 330
CFR
Code of Federal
Regulations 23
civil aviation 62, 163
cockpit20, 45, 63, 72, 75,
109, 118, 126, 127, 128,
129, 131, 139, 177, 199,
200, 202, 215, 257, 259,
263, 266, 268, 271, 275,
279, 284, 285, 287, 300,
303, 312, 314, 317, 336,
365, 369, 375, 387

cockpit designs20, 46, 72, 118, 129, 177

cockpit instrumentation20, 75, 257

cockpit scanning 128

cockpit standardization45, 75

Code of Federal Regulations
CFR 23, 171

cognitive development.....46

commercial4, 17, 23, 24, 28, 29, 43, 54, 57, 58, 82, 88, 89, 90, 95, 103, 104, 108, 110, 115, 120, 126, 127, 135, 137, 141, 148, 156, 159, 173, 179, 186, 193, 199, 201, 207, 215, 237, 238, 239, 240, 241, 246, 249, 251, 270, 292, 293, 295, 296, 299, 309, 316, 327, 330, 336, 337, 349, 350, 357, 368, 370, 432

commercial airline industry 110

commercial aviation17, 28, 82, 88, 89, 103, 120, 148, 159, 179

commercial pilots24, 95, 135

community actors 41

consequences24, 27, 35, 40, 45, 56, 59, 68, 78, 81, 83, 85, 86, 93, 101, 102, 105, 110, 122, 156, 162, 175, 178, 180, 279, 314, 323, 326, 342, 347, 349, 352, 371

Conventional aircraft61, 140

corporate aviation activities 17

Crew resource management
CRM...............................61

crewperson 108

CRM
Crew resource management
..................................61

crop dusting 17

CRT
cathode ray tube 126

cultural values.................35

cultures ..35, 37, 42, 97, 173

curriculum110, 257, 299, 301, 305, 308, 344, 354, 374

curriculums25, 90, 104, 257, 301, 305, 334, 343, 353, 354, 359

data analysis188, 195, 196, 207, 216, 219, 221, 224, 225, 226, 236, 237, 254, 369

data saturation31, 54, 202, 250

decision-making13, 14, 22, 29, 34, 43, 49, 50, 53, 65, 67, 68, 71, 74, 91, 98, 103, 104, 105, 108, 112, 119, 123, 124, 128, 138, 139, 153, 164, 169, 174, 180, 181, 182, 186, 188, 198, 200, 204, 205, 208, 214, 215, 227, 233, 235, 237, 243, 251, 252, 253, 254, 264, 267, 276, 281, 289, 290, 292, 295, 298, 302, 306, 307, 308, 310, 311, 316, 318, 319, 321, 324, 325, 326, 327, 328, 329, 331, 333, 335, 337, 338, 339, 340, 344, 346, 348, 353, 354, 356, 357, 358, 359, 360, 361, 384, 393

decision-making skills14, 29, 104, 281, 356, 357, 359
descriptive case studies 183
descriptive case study29, 182, 188, 236, 329
deterministic beliefs75, 82, 99, 110, 175
dispatchers 61, 64, 146, 159
dissertation 4
dual flight instruction........17
economic drivers 37
embedded units of analysis 29, 54, 185, 186, 188, 190, 197, 199, 201, 235, 236, 241, 250, 251, 254, 292, 293, 302, 316, 327, 330, 331
emerging technologies38, 45, 87, 114, 146, 154, 155, 166, 172, 175, 348, 368
FAA
    Federal Aviation Administration14, 19, 20, 22, 23, 24, 25, 26, 30, 44, 52, 57, 60, 61, 62, 72, 74, 76, 80, 83, 85, 89, 90, 92, 93, 94, 95, 102, 104, 106, 107, 112, 113, 114, 115, 121, 130, 131, 132, 134, 135, 139, 140, 141, 143, 146, 148, 155, 156, 157, 159, 160, 161, 162, 163, 170, 171, 172, 177, 198, 203, 257, 263, 297, 298, 299, 300, 308, 313, 318, 325, 333, 336, 339, 342, 343, 348, 349, 352, 358, 367, 368, 376, 377
face-to-face interviews32, 192, 198, 205, 213, 230, 240
failure of design 38

fatalities17, 20, 23, 27, 47, 56, 59, 67, 73, 78, 88, 102, 123, 129, 177, 178, 180, 183, 185, 259, 261, 283, 315
faulty designs...................90
Federal Aviation Administration
    FAA............14, 20, 367, 377
Federal Aviation Regulations Part 91...151
first-hand experiences50, 54, 181, 226
FITS training 107, 163
Fixed-wing aircraft 61
flight business offices52, 193, 200, 204, 213, 238
flight controls 18
flight deck 19, 132, 143, 367
Flight Deck Human Factor Research Program130, 131
flight information 138, 347
flight logbooks29, 31, 53, 54, 187, 192, 196, 199, 210, 213, 215, 219, 220, 330
flight management systems
    FMS 89, 126
flight maneuvers 118
flight operations66, 72, 82, 105, 159, 259, 284
flight schools52, 193, 200, 204, 238
flight test 17
flight traffic 19
flight training13, 17, 20, 22, 28, 46, 58, 71, 75, 79, 88, 94, 109, 113, 119, 123, 132, 135, 137, 138, 139, 141, 142, 145, 153, 158, 163, 174, 177, 189, 215, 228, 252, 257, 265, 272, 284, 288, 299, 301, 305,

311, 312, 320, 323, 324,
326, 328, 333, 334, 336,
337, 338, 340, 343, 344,
346, 347, 350, 351, 353,
354, 355, 356, 359, 360,
368, 384

flying clubs.......52, 193, 204

flying skills45, 57, 78, 80,
83, 273, 278, 279, 284,
287, 297, 314, 349, 369

FMS
  flight management systems
 126

Future Aviation Advisory
  Committee.........154, 376

GAJSC
  General Aviation Joint
    Steering Committee74,
    75, 107, 108, 140, 301,
    306, 308, 313, 321

GAO
  Government Accountability
    Office17, 18, 20, 21, 23,
    24, 32, 45, 62, 74, 75,
    76, 82, 87, 88, 92, 93,
    96, 102, 107, 108, 112,
    113, 123, 134, 135, 148,
    151, 152, 154, 156, 157,
    159, 161, 177, 214, 215,
    376

general aviation12, 17, 18,
  20, 21, 23, 24, 26, 27, 29,
  40, 41, 44, 45, 47, 49, 51,
  52, 53, 56, 58, 59, 62, 67,
  68, 69, 71, 72, 73, 74, 75,
  78, 81, 83, 86, 88, 90, 91,
  92, 94, 95, 96, 99, 100,
  102, 103, 104, 105, 106,
  107, 108, 109, 111, 112,
  115, 117, 119, 120, 123,
  126, 127, 128, 129, 132,
  134, 135, 137, 138, 139,
  141, 144, 145, 147, 149,
  151, 153, 154, 155, 156,
  157, 159, 161, 162, 166,

167, 169, 170, 171, 173,
174, 177, 178, 180, 181,
183, 185, 186, 187, 189,
190, 191, 192, 197, 198,
199, 200, 201, 202, 203,
204, 206, 208, 210, 211,
213, 214, 215, 216, 221,
227, 228, 229, 230, 233,
235, 237, 238, 239, 240,
242, 243, 249, 252, 257,
259, 261, 262, 264, 266,
270, 278, 281, 282, 285,
295, 297, 299, 300, 304,
306, 308, 310, 311, 312,
313, 316, 320, 322, 323,
326, 327, 328, 329, 330,
333, 335, 337, 338, 339,
340, 342, 343, 345, 346,
348, 350, 351, 352, 354,
355, 356, 358, 359, 361,
391, 432

General Aviation Joint
  Steering Committee
  GAJSC.....74, 107, 306, 368

general aviation
  maintenance...............100

glass cockpits12, 20, 21,
  24, 25, 31, 45, 58, 63, 68,
  73, 75, 91, 96, 103, 107,
  111, 126, 127, 129, 132,
  134, 139, 155, 163, 200,
  215, 237, 238, 243, 252,
  256, 258, 264, 266, 272,
  299, 300, 301, 303, 307,
  312, 324, 344, 372, 393

global positioning systems
  GPS 63, 92, 271

Government Accountability
  Office
  GAO32, 62, 113, 214, 342,
    376

government officials.........48

government oversight20,

22, 40, 43, 56, 58, 67, 72,
73, 140, 156, 159, 177
GPS
global positioning systems
.....................................92
ground controllers.............42
guidelines4, 65, 173, 219,
369
hazardous impacts 93
hazards27, 65, 72, 73, 84,
90, 115, 117, 142, 145,
150, 172, 178, 329, 351
holistic training.......140, 322
human factors24, 40, 44,
46, 57, 64, 67, 71, 72, 75,
88, 90, 92, 112, 115, 117,
118, 119, 120, 122, 129,
131, 132, 134, 142, 143,
145, 174, 214, 279, 284,
301, 314, 338, 348, 361,
364, 369, 373, 377
Human Factors Analysis
and Classification
System 119
human factors training ...120
human interface...............76
human-machine interaction
.................40, 47, 86, 119
individual actors...............41
industrialization..........35, 36
informed consent204, 212,
217, 231, 332, 373, 377,
392
informed consent form212,
217, 231, 239
innovations19, 26, 35, 37,
38, 44, 47, 50, 58, 66, 71,
73, 75, 77, 78, 80, 82, 83,
84, 86, 88, 92, 95, 97, 98,
99, 102, 103, 105, 111,
113, 114, 116, 118, 120,
122, 130, 134, 138, 151,
152, 156, 159, 160, 161,
162, 165, 166, 168, 173,

177, 181, 185, 253, 254,
265, 277, 286, 294, 295,
309, 342, 345, 346, 347,
349, 351, 355, 360, 432
institutional practices39,
162, 172, 328
instrument........................54
instrument-rated 29
integrated systems 42
integrating technologies44,
73
integration of technological
developments43, 158,
172
International Civil Aviation
Organization 120
Interview Question Guide
205, 206, 207, 209, 236,
237, 242, 243, 250, 251,
394
LinkedIn 2
liquid crystal displays.....126
machines 42
maintenance workers 121
member check196, 218,
227, 250
micro-level transformation
.............................34, 44
military23, 36, 43, 58, 88,
103, 104, 108, 127, 141,
156, 228, 279, 314, 330,
350, 365, 431
military aviation................17
NASA
National Aeronautics and
Space Administration
..........113, 114, 126, 366
National Aeronautics and
Space Administration
NASA 60, 113
national airspace system42,
61, 64, 106, 113, 146,
167, 171, 376

National Transportation
Safety Board
NTSB 22, 64, 256, 372
new technologies20, 26, 28,
34, 38, 39, 42, 43, 45, 56,
68, 69, 71, 72, 76, 85, 86,
88, 92, 94, 95, 96, 97, 98,
105, 109, 114, 116, 131,
134, 143, 144, 145, 146,
150, 159, 164, 166, 172,
174, 179, 184, 278, 279,
287, 314, 323, 328, 334,
335, 343, 345, 346, 352,
355, 359, 385, 432
Next Generation of Air
Transportation System
NextGen..........................105
NextGen
    Next Generation of Air
    Transportation System
    105, 106, 146, 147, 365,
    371, 374
NTSB
    National Transportation
    Safety Board17, 20, 22,
    23, 27, 32, 45, 46, 47,
    56, 57, 60, 63, 64, 65,
    67, 68, 73, 74, 75, 76,
    78, 79, 80, 82, 83, 88,
    90, 92, 96, 103, 110,
    112, 126, 127, 128, 129,
    134, 135, 139, 140, 141,
    175, 177, 178, 180, 184,
    185, 214, 262, 298, 299,
    300, 301, 303, 305, 308,
    310, 320, 322, 333, 339,
    342, 344, 345, 346, 351,
    372
operational scenarios 104
operator training 42, 43
organizational culture26, 34,
    75, 157, 162
orientation training ...28, 179
paradigms20, 26, 40, 80,
    103, 104, 113, 137, 139,

142, 169, 174, 177, 301,
311, 315, 319, 337, 340,
356
philosophical beliefs ..48, 59
photography....................17
pilot awareness...............19
pilot behavior 145
pilot error18, 27, 57, 65, 68,
    73, 104, 112, 120, 178,
    283, 342, 372
piloting skills 94, 127, 136
pilots13, 17, 19, 21, 22, 23,
    24, 25, 27, 29, 42, 44, 49,
    50, 51, 52, 54, 57, 58, 61,
    62, 63, 64, 65, 67, 71, 72,
    73, 74, 75, 81, 83, 87, 88,
    89, 90, 91, 92, 94, 95, 96,
    102, 104, 105, 106, 107,
    108, 109, 111, 112, 117,
    118, 121, 123, 127, 128,
    130, 133, 135, 136, 138,
    139, 140, 141, 142, 144,
    145, 146, 151, 153, 158,
    159, 161, 163, 165, 169,
    174, 177, 178, 179, 181,
    182, 184, 186, 187, 188,
    189, 191, 192, 197, 198,
    199, 200, 201, 202, 203,
    205, 206, 207, 210, 211,
    213, 215, 216, 220, 227,
    228, 229, 230, 233, 235,
    237, 238, 239, 241, 242,
    243, 244, 246, 249, 251,
    252, 253, 257, 258, 259,
    260, 261, 262, 263, 265,
    267, 269, 270, 273, 275,
    276, 277, 279, 280, 281,
    282, 284, 285, 286, 288,
    289, 291, 293, 294, 295,
    296, 297, 298, 299, 300,
    301, 302, 303, 304, 305,
    306, 307, 308, 309, 311,
    313, 314, 315, 316, 317,

319, 320, 321, 324, 325,
326, 327, 328, 329, 330,
333, 335, 336, 338, 339,
340, 341, 343, 344, 346,
347, 348, 349, 351, 353,
354, 355, 356, 358, 359,
360, 361, 364, 365, 366,
369, 371, 372, 374, 384,
392, 393
pilots' judgment..............104
political factors.................37
politics........36, 97, 166, 173
predictive analysis45, 79,
122, 133, 146, 348
private pilots29, 186, 237,
244, 249, 251, 261, 263,
266, 268, 295, 296, 309,
316, 327, 330
problem solving skills.....104
purposeful sample30, 51,
202
purposeful sampling52, 198,
200, 201, 203, 210, 238
qualifications31, 186, 199,
206, 215, 331
qualitative research30, 49,
53, 180, 194, 196, 197,
201, 206, 220, 363, 369,
370, 371, 374, 375
recreational17, 23, 43, 82,
87, 264
recreational flying 17
regulations12, 23, 25, 27,
37, 40, 58, 59, 64, 65, 67,
72, 88, 89, 94, 114, 115,
118, 121, 134, 138, 143,
144, 152, 155, 162, 166,
170, 173, 175, 178, 231,
261, 264, 298, 316, 326,
328, 329, 335, 342, 352,
353, 358, 369
regulatory oversight14, 27,
42, 57, 72, 80, 143, 170,
171, 178, 180, 316, 323,
349, 352
researchers......................48
resource management13,
45, 58, 61, 66, 68, 74,
103, 105, 108, 109, 119,
139, 174, 264, 276, 282,
292, 295, 298, 302, 307,
308, 316, 317, 318, 322,
324, 325, 328, 333, 334,
339, 340, 345, 347, 348,
354, 360
retooling 153
risk management22, 65, 75,
91, 105, 149, 170, 298,
318, 348, 354, 367
safeguard24, 26, 76, 86, 93,
146, 232
safety12, 13, 14, 20, 24, 25,
27, 37, 45, 47, 49, 50, 53,
57, 58, 61, 65, 66, 67, 71,
73, 74, 75, 78, 79, 82, 83,
86, 87, 88, 89, 92, 93, 95,
106, 107, 109, 111, 114,
115, 116, 118, 120, 122,
131, 134, 135, 140, 145,
146, 147, 149, 151, 153,
154, 158, 160, 163, 165,
170, 172, 177, 178, 179,
182, 186, 188, 198, 200,
204, 214, 215, 233, 235,
237, 238, 251, 253, 254,
256, 258, 259, 262, 264,
271, 274, 275, 277, 279,
287, 290, 292, 294, 298,
299, 304, 306, 308, 309,
310, 311, 312, 313, 314,
315, 316, 318, 320, 321,
323, 324, 325, 326, 327,
328, 329, 331, 332, 334,
335, 337, 338, 339, 342,
343, 347, 349, 350, 351,
352, 354, 355, 356, 358,

359, 360, 361, 365, 366, 367, 370, 371, 372, 373, 374, 376, 385, 392, 393, 432

safety experts 20, 45, 47

safety initiatives28, 46, 88, 160, 179

safety management102, 149, 215, 371

Safety Management System SMS 83

Safety regulations............65

safety risk management SMR..............................150

Scenario-based training...65

scenarios 65, 95, 105

scientific research12, 57, 73, 79, 81, 85, 99, 127, 129, 154, 158, 165, 185

search and rescue 17

simulators 118, 266

single pilot resource management108, 109, 293, 307, 308, 317, 318, 322, 325, 345

Single-pilot resource management 66

situation awareness13, 28, 78, 80, 83, 91, 111, 118, 121, 124, 126, 130, 179, 254, 295, 327, 333, 364, 370

Situational awareness66, 294

SMR
    safety risk management.150

SMS
    Safety Management System 83, 84, 87, 88, 133, 147, 148, 149, 150, 152, 160, 161, 367

snowball sampling13, 30, 51, 52, 196, 202, 203, 363

social actors.....................37

social tenets.....................40

socio-technical system42, 43, 120, 122, 134, 147, 375

socio-technical systems41, 43, 122, 132, 155, 167, 174

software upgrades 90

system designers...........124

system designs12, 27, 90, 178, 329, 358

system processes............27

system redundancy 19

TAA
    technologically advanced aircraft12, 13, 14, 18, 19, 20, 21, 22, 23, 24, 25, 27, 28, 29, 30, 31, 32, 33, 40, 41, 44, 45, 46, 47, 48, 49, 50, 51, 52, 53, 54, 56, 57, 58, 59, 61, 63, 67, 68, 71, 72, 73, 74, 75, 76, 78, 80, 81, 82, 86, 88, 89, 90, 91, 93, 94, 95, 96, 99, 101, 103, 104, 107, 108, 109, 110, 111, 112, 113, 118, 126, 127, 128, 129, 132, 133, 134, 135, 136, 137, 138, 139, 140, 141, 142, 145, 153, 157, 159, 162, 163, 166, 169, 170, 171, 174, 175, 177, 178, 179, 180, 181, 182, 183, 184, 185, 186, 187, 188, 189, 190, 191, 192, 193, 197, 198, 199, 200, 201, 202, 203, 204, 205, 206, 208, 210, 211, 213, 214, 215, 216, 219, 221, 227, 228, 229, 230, 233, 234, 235, 237, 238, 239, 240, 241, 242, 243, 244, 245, 246, 247, 248, 249, 250, 251, 252, 253, 258, 260, 261, 262, 263, 264, 265,

266, 267, 268, 269, 270,
271, 272, 273, 274, 276,
277, 278, 279, 280, 281,
282, 283, 284, 285, 286,
287, 289, 290, 291, 293,
294, 295, 296, 297, 298,
299, 300, 301, 302, 303,
304, 305, 306, 307, 308,
309, 310, 311, 312, 313,
314, 315, 316, 317, 318,
319, 320, 321, 322, 323,
324, 326, 327, 328, 329,
330, 331, 333, 334, 335,
336, 337, 338, 339, 340,
341, 342, 343, 344, 345,
346, 347, 348, 350, 351,
353, 354, 355, 356, 357,
358, 359, 360, 361, 383,
384, 385

technical determinism
    theory 235
technological actors.........41
technological determinism
    33, 34, 36, 38, 40, 41, 45,
    47, 56, 66, 69, 77, 86, 89,
    97, 98, 106, 169, 185,
    226, 236, 295, 323, 326,
    328, 340, 345, 355, 357,
    359, 372, 376, 391
technological infrastructure
   160
technological innovations72,
    77, 84, 97, 120, 167, 168,
    169, 286, 309
technologically advanced
    aircraft
    TAA...............................12
technologists47, 48, 80, 85,
    86, 92, 185, 189
technology failures...........45
terrain 19, 78
themes13, 31, 49, 54, 55,
    187, 189, 194, 201, 207,
    216, 220, 221, 223, 224,

226, 233, 236, 237, 251,
254, 256, 257, 258, 259,
292, 294, 295, 296, 302,
309, 316, 327, 332
thesis 4
training13, 14, 18, 19, 22,
    23, 24, 26, 27, 40, 44, 47,
    56, 58, 59, 65, 67, 68, 72,
    75, 79, 80, 83, 84, 86, 88,
    89, 90, 91, 95, 96, 103,
    104, 106, 107, 109, 112,
    113, 120, 124, 125, 129,
    132, 134, 135, 137, 138,
    139, 141, 142, 143, 145,
    159, 163, 174, 177, 178,
    180, 191, 214, 215, 235,
    242, 252, 253, 254, 257,
    258, 259, 261, 262, 264,
    265, 268, 271, 274, 275,
    277, 278, 279, 280, 284,
    288, 290, 292, 293, 295,
    296, 297, 298, 299, 301,
    302, 304, 305, 306, 308,
    309, 310, 312, 313, 314,
    315, 316, 317, 318, 319,
    321, 322, 324, 325, 327,
    332, 333, 336, 338, 339,
    340, 342, 343, 345, 346,
    347, 348, 350, 352, 353,
    354, 355, 357, 359, 360,
    361, 364, 366, 368, 369,
    370, 371, 373, 374, 375,
    376, 384
training deficiency............73
training requirements90,
    138, 139, 163, 263
van Kaam 54, 219, 226, 251
van Kaam model......54, 220
Vulnerability analysis 166
weather19, 62, 63, 64, 76,
    78, 130, 285, 341
work overload 118

CURRICULUM VITAE

Calvin Nobles, Ph.D.

PSC 482 Box 2666, FPO, AP 96362
cn8972@gmail.com
https://jp.linkedin.com/in/calvinnobles
(443) 951-5996

STATEMENT OF TEACHING PHILOSOPHY

As an instructor, I understand the importance of fostering an environment that creates dialogue, participation, and teamwork to increase the effectiveness of the learning atmosphere. As an instructor, I employ a variety of strategies to motivate and assist students in developing their ability to cultivate knowledge and professional competencies. I deliver the curriculum in a manner that is informative and inspiring, incorporating current events, and "lessons learned" from literature and professional experiences to supplement course material to provide an enriched and meaningful learning experience to students. I use themes, integrated units, projects, group and individual assignments, and online collaboration to enhance the learning environment and to challenge the students.

One of the most important concepts I impart to students is the intrinsic value of being a lifelong learner. I encourage students to go beyond the minimum to increase their intellectual capacity and evaluate how course content is applicable to one's current career. An invaluable lesson I learned was the importance of critical thinking, which is invaluable in one's professional or personal life. All students should depart their institutions armed with the ability to think critically through the most complex problems.

A cornerstone of my teaching philosophy is to help students develop their critical thinking and higher learning abilities. Realizing that students possess different learning styles and abilities, I present the class materials in a variety of formats, using lectures, articles, visual learning aids, organized notes, charts, and diagrams to ensure each student is capable of comprehending the information. My goal is to make sure each student is equipped with the information presented in the course to aid his or her efforts in pursuing educational, professional, and intellectual objectives.

EDUCATION

- 2015, Ph.D. in Business Administration, Management and Engineering Technology,

Northcentral University, Prescott Valley, AZ;
Dissertation Title: Exploring Pilots' Experiences of
Integrating Technologically Advanced Aircraft
Within General Aviation: A Case Study

- 2010, Masters Of Arts, Military Operational Art
  And Science, AIR University, Maxwell Air Force
  Base, AL; Thesis Topic: A Continuity of Failure: A
  Comparison Between Vietnam and Afghanistan
- 2009, MBA In Management And Engineering
  Technology Northcentral University, Prescott
  Valley, AZ
- 2003, Masters Of Aeronautical Science, Embry
  Riddle Aeronautical University, Daytona Beach,
  FL; Thesis Topic: control flight into terrain in
  general aviation
- 1998, Bachelor of Science in Management, Park
  University, Parksville, MO

TEACHING EXPERIENCE

06/15 – present, Adjunct Faculty
Central Texas College, Camp Foster, Okinawa
Japan

- Instructor for undergraduate level business
  courses. Taught the following courses during the
  summer semester:

Principles of Marketing MRKG 1311
This course provides the basic marketing functions;

identification of consumer and organizational needs; explanation of economic, environmental, psychological, sociological, and global issues; and description and analysis of the importance of marketing research.

Supervision BMGT 1301
This course examines the roles of the supervisor and managerial functions as applied to leadership, counseling, motivation, and human skills are examined.

Approved to teach the following courses:

Business/Office Automation BMGT 1325
This course examines systems, procedures, and practices related to organizing and planning office work, supervising employees' performance, and exercising leadership skills.

Information Technology Security for Homeland Security Specialists HMSY 1370
This course covers the basics of information technology security. Topics covered are: the uses of cyber-crime by terrorist organizations and their impact on our nation's information-based infrastructure, government, corporate, and private institutions and citizens; how to protect data and infrastructure from cyber-crimes and electronic terrorism. Desktop computer, organizational infrastructure, communications infrastructure, and network security

will be examined. The course includes a study of the uses of computer forensics and methods to defend against cyber-attacks. It will examine applications with proven success and apply to real-life scenarios.

Small Business Management BUSG 2309
This course examines starting, operating and growing a small business. Topics include facts about small business, essential management skills, how to prepare a business plan, accounting strategies, financial needs, staffing, marketing strategies, and legal issues.

Business Law BUSI 2301
The course provides the student with foundational information about the U.S. legal system and dispute resolution, and their impact on business. The major content areas will include general principles of law, the relationship of business and the U.S. Constitution, state and federal legal systems, the relationship between law and ethics, contracts, sales, torts, agency law, intellectual property, and business law in the global context.

Introduction to Ethics PHIL 2306
This course examines the systematic evaluation of classical and/or contemporary ethical theories concerning the good life, human conduct in society, morals, and standards of value.

Principles of Management BMGT 1327
Examines concepts, terminology, principles, theory, and issues that are in the field of management. This course focuses on the challenges of change and management's response to change, the diversity of management methods, and managing strategies for the future. As a seminar, this course uses peer teaching and learning approaches, involves group learning experiences in a team environment, requires practical application of concepts, and includes research and case studies. This course culminates the associate's degree of management.

Management Problems and Application I and II BMGT 2370 and 2371
This course emphasizes management decision-making skills that are necessary for the modern manager to successfully deal with resource problems in the work place and career field. The course is designed for students desiring to further their management training by applied learning techniques utilizing decision-making case studies, computer-assisted models, or other instructional techniques. It serves as a program Capstone for those students whose situation precludes an internship.

01/13 – 12/13, Adjunct Faculty - Satellite Program Army Command & Staff College, Fort Gordon, GA

- Taught a graduate level course on naval and maritime operations:

Naval and Maritime Operations
This course provides a graduate level approach to naval and maritime strategies, and amphibious operations. This course introduces students to naval command and control and the employment of naval forces in support of Geographical Combatant Commanders. Students will learn how naval and maritime capabilities are leveraged in a joint construct or in the maritime domain.

07/11 – 12/13, Naval Instructor
Navy Information Operations Command Georgia
US Navy, Fort Gordon, GA

* Provided undergraduate and graduate level training to Naval Officers. Taught the following courses:

National Security Policy
This course introduces the issues and institutions of national security policy. Students will gain an appreciation of strategic thought and strategy formulation, the ability to assess national security issues and threats, and an understanding of the political and military institutions involved in the making and execution of national security policy.

Business Communications
This examines application of business communication principles through creation of effective business

documents and oral presentations. Includes study and application of team communication and use of technology to facilitate the communication process.

Personnel Management and Administration
This course examines how people work in government and nonprofit settings, especially for those who direct and manage others or plan to do so in the future. The course attempts to teach managers and supervisors the laws, procedures, and techniques of public personnel management. The course also covers what motivates workers and how to match organizational needs with the individual talents and interests of employees. The course explains the prescriptions and the restrictions of personnel management that applies to government and nonprofits, and the policy rationale for this framework. In short, the scope of the course includes concern for both the generic issues of human behavior in organizational settings and the specific needs of public sector accountability in a democracy.

Cyber Security Oversight
This course examines mid-level security practitioners how to engage all functional levels within the enterprise to deliver information system security. To this end, the course addresses a range of topics, each of which is vital to securing the modern enterprise. These topics include plans and policies, enterprise roles, security metrics, risk management, standards and regulations, physical security, and

business continuity. Each piece of the puzzle must be in place for the enterprise to achieve its security goals; adversaries will invariably find and exploit weak links.

Organizational Behavior

This course examines the logical and rational design of organizations; emergent behavior in the individual and the group, including interaction and effect on the organization.

Organizational Change

This course examines real-world issues organizations face when they expand internationally, merge, reshape the structure, new or modify operations and processes, introduce new technologies, respond to government regulations, react to competition and respond to changing customer needs. A common theme in these different challenges and undertakings is the need to change. Change is becoming a regular feature of organizational life. This course introduces students to the theory and practice of organizational change. The challenges faced by individuals in leadership and followership positions in creating, or responding to changes will be addressed.

Intelligence Analysis

This course focuses on building on the Seminar on Issues in Intelligence Analysis. Students will complete and present solutions for team-based intelligence community or competitive intelligence IA projects.

Students will produce written and oral technical reports/briefs of their results.

Terrorism
This course examines the study of terrorism and reasons why America is a terrorist target. Includes methods of combating domestic and international terrorism, terrorist operations, cyber-terrorism, narco-terrorism, the mind of the terrorist, and organized crime's impact on terrorism.

Innovation Management
The course aims to equip management students with an understanding of the main issues in the management of innovation and an appreciation of the relevant skills needed to manage innovation at both strategic and operational levels. It provides evidence of different approaches based on real-world examples and experiences of leading organizations from around the world.

Business Strategy
This course examines strategy involves the coordination and integration of the efforts within the different functional areas of an organization for dealing with an uncertain future. This comprises formulating a business strategy for each individual unit of the firm, formulating a corporate strategy, and implementing these strategies. Strategy formulation involves understanding the business the firm is in, determining how to position the strategic unit within

this business environment, and developing the capabilities to compete, but also to cooperate, in this environment. Therefore, strategy at the same time coordinates and integrates the individual functional strategies such as finance, human capital, and tasks prioritization.

PROFESSIONAL EXPERIENCE

**02/14 – present, Deputy Assistant
Chief of Staff Intelligence
Cryptologic Resource Coordinator (Military)
US Navy, Okinawa, Japan**

- Led and coordinated the strategic planning, management, and execution of all cryptologic operations in support of amphibious operations for Commander Seventh Fleet.
- Responsible for overseeing all functions of cryptologic and signals intelligence operations to include analysis, staffing, management, and training.
- Responsible for developing and aligning cryptologic efforts to support intelligence needs for an Expeditionary Strike Group.
- Coordinated national, theater, and tactical level electronic warfare, cyber, and signals intelligence to support indications and warning and threat warnings.
- Managed the fusion of cryptology and signals intelligence into intelligence products

- Coordinate cyber security and cyber expertise for deployed naval ships to enhance computer network defense measures and information assurance to maintain network continuity naval information operation command Georgia, fort Gordon, GA

2012 – 2013, Department Head

- Led and managed 510 personnel in supporting over 30 intelligence mission areas.
- Managed and directed over 8,000 administrative requirements. Reduced the administrative completion dates by 33 percent by streamlining processes, empowering divisional, and branch level managers.
- Coordinated an administrative reorganization to align personnel with 12 functional areas to reduce operational cost, enhance collaboration, improve operational efficiency, and reduce operational latency by 27 percent.
- Member of the Cloud Computing Training Team and assisted with migrating 1700 personnel from traditional network infrastructure to thin clients and the cloud network. Provided in-depth training on using cloud computing and the thin client network.
- Directed strategic, operational, and technical level training programs for 510 personnel to improve operational efficiency, reduce rework, and to enhance professional development.
- Led the development of the Cyber Operations

Department to address maritime cyber requirements and support for operational commanders. Assisted with the development of the information technology instructional and cyber requirements. Trained 30 naval officers on information technology and cyber requirements to increase support to operational commanders.

- Authored and implemented strategic plans and policies to improve organizational health, operational efficiency, and to strengthened long-term partnerships.
- Provided signals intelligence to support Homeland Defense. Responsible for managing requirements to enforce homeland security through sensor management, indications and warning, and threat warning.

2011 – 2012, Deputy Chief
Signals Intelligence, Strategy, and Governance
National Security Agency (NSA)
Central Security Service, Fort Gordon, GA

- Managed a team of 53 technical experts in developing solutions for a multitude of consumers. Responsible for developing strategies to implement new technologies to enhance intelligence analysis
- Directed and managed strategies for the development of new technologies to enable consumers to increase their technical capabilities.
- Coordinated and trained 53 technical specialists in

five competent areas to work with customers on designing new technological capabilities based the consumers' organizational requirements. Implemented training strategies to achieve corporate training objectives by leveraging new technologies, tradecraft, and new methodologies

- Managed a team of computer engineers, computer programmers, software engineers, and intelligence analysts to develop automated capabilities to enhance existing processes and procedures. These automated capabilities reduced project completion dates by 12 weeks and reduced the required manpower by 75 percent.
- Led the technological integration efforts for automated functions to that reduced analytical workload by 65%.
- Served as a member of the Executive Training Committee. Responsible for implementing new training courses to enhance the workforce knowledge of new technologies and future capabilities.

2010 – 2011, Operations Team Chief
European Command
US Navy, Stuttgart, Germany

- Supervised a team of 10 individuals specialized in coordinating operations, administrative support, situational awareness, and organizing efforts between nine sub-organization
- Responsible for facilitating and training personnel

for 10 functional positions to support joint military operations

- Coordinated and trained all specialized personnel on the administrative, technological, and customer support processes to increase awareness and communication between component commands and the theater commander

2007 – 2009, Division Officer
US Navy, USS Theodore Roosevelt (CVN 71)

- Supervised and trained 132 personnel in Hangar Bay Operations in support of Operation Enduring Freedom
- As a member of the Air Department Training Team responsible for training 750 personnel in flight deck operations, flight deck firefighting, and flight deck emergencies
- Assisted with the safe catapult launch and recovery of over 15,000 aircraft while managing two catapult systems and an arresting gear team
- As a Helicopter Control Officer directly responsible for the safe launch and recovery of all helicopters, whiling supervising personnel involved in flight deck operations on an aircraft carrier and assisted with launch preparation on a daily basis

2004 - 2007. Branch Chief
National Security Agency
Central Security Service
US Navy, Fort Meade, MD

- Provided in-depth training to 152 personnel while in a combat environment and responsible for re-engineering tactical systems to improve tactical targeting and intelligence
- Advised five Brigade Combat Teams on intelligence support and enhanced coordination between national level intelligence organizations and tactical units
- Established new intelligence methodologies to reduce reporting requirements

PUBLICATIONS

Nobles, C. (2015). Exploring Pilots' Experiences of Integrating Technologically Advanced Aircraft Within General Aviation: A Case Study. Virginia Beach, VA: D. Boyer Consulting.

Nobles, C. (2015). Exploring Pilots' Experiences of Integrating Technologically Advanced Aircraft Within General Aviation: A Case Study. ProQuest. Ann Arbor, MI.

PROFESSIONAL COMMITTEES

2012 – 2013 – New Employee Orientation Briefing
Team, Fort Gordon, GA

Provided in-depth briefs to new employees on
maritime operations, signals intelligence, information
warfare, electronic warfare, and cyber warfare.

2011 – 2102 – Executive Training Committee, Fort
Gordon, GA. Provided in-depth and holistic
executive-level support and guidance on developing
training requirements for intelligence analysts,
computer engineers, and computer scientists.
Responsible for developing strategic objectives and
requirements to maintain pace with advancing
technology to improve analytical capabilities.

2005 – 2007– DOD Interagency. Joint Military
Targeting Committee, Washington, DC.
Provided subject matter expertise to codify
techniques, tactics, and procedures for
developing kinetic and non-kinetic targets to assist
operational commanders with the
ability to expeditiously construct targeting data
through a vetted process.

2003 – 2004 – U.S. Navy. Human Factors Council.
Fleet Air Reconnaissance Squadron Four, Tinker
AFB, OK. Council Member. Participated in over 50

Aviation Human Factors Council meetings to address adverse air crewmen's issues and condition by providing in-depth resolutions to the Squadron Commander to maintain air operations effectiveness.

PROFESSIONAL PRESENTATIONS & CONFERENCES

Nobles, C. (2015, July). Impacts of Integrating Technologically Advanced Aircraft in General Aviation. Presented at Kadena Flying Club, Okinawa, Japan.

Nobles, C. (2015, April). Amphibious Operations in Seventh Fleet Area of Operations. Presented at United Nations Command-Rear Conference, Okinawa, Japan.

Nobles, C. (2015, March). Intelligence Capabilities Of A Forward Deployed Expeditionary Strike Group. Presented at 1st Annual Okinawa-wide Intelligence Symposium. Kadena Air Force Base, Okinawa, Japan.

Nobles, C. (2014, April). Commander Seventh Fleet Commander's Intelligence Waterfront Conference. Expeditionary Strike Group and Amphibious Operations. Yokosuka, Japan.

Nobles, C. (2012, September). Navy Information Operations Command Georgia. Fort Gordon, GA. Naval information operations and information warfare capabilities.

Nobles, C. (2011, August) Signals Intelligence Development, Strategy, and Governance. Fort Gordon, GA. Developing new capabilities to support analytical efforts and implementing new tradecrafts to enhance analysis and production.

Nobles, C. (2007, April). Defense Science Board Task Force on Time Critical Strike from Strategic Standoff. Sensitive Targeting Directed Energy Solutions for Offensive Strike of Time Critical Targets. Washington, DC.

Nobles, C. (2006, March). Joint Military Targeting Conference. Developing Capabilities to Provide Signals Intelligence in Support of Information Operations. Washington, DC.

PROFESSIONAL EDUCATION & HONORS

- Inducted into Delta Mu Delta International Honor Society-2012, for outstanding academic achievement.

PROFESSIONAL ORGANIZATIONS

- Aircraft Owners and Pilot Association 2010-Present
- University Aviation Association 2009-Present
- National Naval Officers' Association 2003-Present
- Omega Psi Phi Fraternity Incorporated 2010-Present
- Delta Mu Delta, International Honor Society for Business 2012-Present
- Kadena Flying Club (Okinawa, Japan) 2014-Present

COMPETENTENCY SUBJECT AREAS

- Business Administration
- Information Technology
- Cybersecurity
- Aeronautical Science
- Physics
- Mathematics
- Military Science

- Aviation Management
- Aviation Operations
- Organizational Leadership
- Organizational Behavior
- Strategic Planning
- Human Resources Management
- EEOC

- Business Intelligence
- International Management
- Financial Management
- Technological Innovation
- Leading Change
- Business Communications
- Organization Administration

COMPETENTENCY SUBJECT AREAS

•Aviation Safety

•Accident Investigation

•Flight Planning

•Meteorology

•Strategic Development

•Management

•National Security Planning

•National Security Policy

•Military Planning

•Military Operations

•Military Intelligence

•Military Leadership

•Human Factors in Aviation

•Crew Resource Management

•Single Pilot Resource

•Leadership

•Logistics

BUSINESS AND TECHNICAL
TRAINING AND CERTIFICATIONS

- Commercial Pilot, Multi-engine Land, Single Engine Land; Instrument Rated
- Private Pilot, Multi-engine Land, Single Engine Land; Instrument Rated
- Helicopter Control Officer
- Arresting Gear Officer
- Catapult Launching Officer
- Aviation Safety Officer
- Mission Commander
- Airborne Communication Officer
- Information Warfare Officer

RESIDENCIES AND COLLOQUIA

Air University, Air Command and Staff College. Maxwell Air Force Base, AL. Attended in residence Joint Professional Military Education and participated in the 10 month Stabilization Operations elective and provided comprehensive guidance and participation in a joint military training exercise.

AWARDS AND HONORS

- Delta Mu Delta, International Honor Society for Business
- Meritorious Service Medal
- Joint Service Commendation Medal
- Navy and Marine Corps Commendation Medal
- Army Commendation Medal
- Joint Service Achievement Medal
- Navy and Marine Corps Achievement Medal
- Combat Action Ribbon
- Global War on Terrorism Expeditionary Medal
- Global War on Terrorism Service Medal
- Iraqi Campaign Medal

COMMUNITY SERVICE & LEADERSHIP

- United States Naval Officer, 1992 – present
- University Aviation Association, 2009 – present
- Aircraft Owners and Pilots Association, 2007 – present
- National Naval Officers Association, 2003 – present
- Omega Phi Psi Fraternity Incorporated, 2010 – present
- Delta Mu Delta, International Honor Society for Business, 2012 – present

ABOUT THE AUTHOR

Dr. Calvin Nobles is from Mount Vernon, GA and is currently serving in the United States Navy. Dr. Nobles earned a B.S. in Management, a Masters in Aeronautical Science, a Masters in Military Operational Art and Strategy, a Masters of Business Administration, and Ph.D. in Business Administration.

Dr. Nobles' professional works consist of leadership, directing, and project management in aviation, cryptology, operations, and military planning areas of expertise. He is a commercial rated pilot and general aviation enthusiast. Dr. Nobles is committed to improving general aviation safety; hence, the reason for this book.

His research focus areas are: technology implementation, technological discontinuities, disruptive innovations, and the adverse implications caused by the continuous integration of new technologies into general aviation. Dr. Nobles is a Life Member of the Omega Psi Phi Fraternity Incorporated.

Contact the author at cn8972@gmail.com if you wish to discuss content or to provide recommendations for future research studies.

Connect on Social Media With Dr. Nobles

https://jp.linkedin.com/in/calvinnobles

ABOUT THE BOOK

A must read for researchers and practitioners with a personal and professional desire to improve general aviation safety. As a general aviation enthusiast and researcher, the author wants to ensure general aviation remains a viable and flourishing sector of the aviation industry. This book highlights a growing phenomenon of integrating innovative technologies in general aviation. The goal is to ensure technologists, researchers and the aviation industry ameliorate collaborative endeavors to counterbalance the pros and cons of integrating technologies in general aviation to decrease the number of accidents and fatalities.

Technologically advanced aircraft are fascinating and pilots with the proper training are able to capitalize on flying advanced aircraft with comparable technologies that are analogous to commercial and military aircraft. This book expounds on general aviation pilots' concerns with integrating glass cockpits in general aviation. The research within corroborates a previous study that technologically advanced aircraft have not improved general aviation safety.

At issue are training, safety, proficiency and competency, and aeronautical decision-making implications. General aviation remains under researched when compared to commercial and military aviation sectors. This book provides recommendations for improving general aviation safety by eliminating the after-effects of innovations in general aviation.

Virginia Beach, VA 23464

http://dboyerconsulting.com
Dawn@DBoyerConsulting.com

KEY BOOK INDEX TERMS: accidents, adaptable automation, aeronautical decision-making, aeronautical examination, air traffic control, air transport industry, aircraft, analog gauges, assessments, Automation management, autopilot, aviation, aviation leaders, Aviation Safety Officer (ASO), Aviation Safety Program (ASP), avionics, certified flight instructors, cockpit instrumentation, cockpit scanning, cockpit standardization, Code of Federal Regulations (CFR), commercial aviation, Crew Resource Management (CRM), curriculum, data analysis, decision-making skills, dispatchers, Federal Aviation Administration (FAA), FITS training, Fixed-wing aircraft, flight management systems (FMS), flight schools, flight training, Future Aviation Advisory Committee (FAAC), General Aviation Joint Steering Committee (GAJSC), glass cockpits, global positioning systems (GPS), Government Accountability Office (GAO), ground controllers, Human Factors Analysis and Classification System, International Civil Aviation Organization (ICAO), National Aeronautics and Space Administration (NASA), National Transportation Safety Board (NTSB), Next Generation of Air Transportation System (NextGen), pilots, Safety Management System (SMS), Single-pilot resource management, snowball sampling, technologically advanced aircraft (TAA), training deficiency, van Kaam model

www.ingramcontent.com/pod-product-compliance
Lightning Source LLC
Chambersburg PA
CBHW051849170526
45168CB00001B/31